Policy and Practice in the Management of Tropical Watersheds

WINROCK DEVELOPMENT-ORIENTED
LITERATURE SERIES
Steven A. Breth, series editor

Policy and Practice in the Management of Tropical Watersheds was prepared under the auspices of Winrock International Institute for Agricultural Development.

ALSO IN THIS SERIES

Rice in the Tropics: A Guide to the Development of National Programs, Robert F. Chandler, Jr.

Small Farm Development: Understanding and Improving Farming Systems in the Humid Tropics, Richard R. Harwood

Successful Seed Programs: A Planning and Management Guide, Johnson E. Douglas

Tomatoes in the Tropics, Ruben L. Villareal

Wheat in the Third World, Haldore Hanson, Norman E. Borlaug, and R. Glenn Anderson

Cassava: New Potential for a Neglected Crop, James H. Cock

Potatoes: Production, Marketing, and Programs for Developing Countries, Douglas Horton

Policy and Practice in the Management of Tropical Watersheds

H. C. Pereira

with illustrations by the author

Routledge
Taylor & Francis Group
LONDON AND NEW YORK

First published 1989 by Westview Press

Published 2019 by Routledge
52 Vanderbilt Avenue, New York, NY 10017
2 Park Square, Milton Park, Abingdon, Oxon OX14 4RN

Routledge is an imprint of the Taylor & Francis Group, an informa business

Library of Congress Cataloging-in-Publication Data
Pereira, H. C. (Herbert Charles), 1913–
Policy and practice in the management of tropical watersheds.
(Winrock development-oriented literature series)
Bibliography: p.
Includes index.
1. Watershed management—Tropics. 2. Agriculture—Tropics. I. Title. II. Series.
TC526.5.P47 1989 333.91'15'0913 88-33794

ISBN 13: 978-0-367-28317-9 (hbk)

Contents

Illustrations

Tables

Figures

Photos

Acknowledgments

I thank Dr. K.G. Tejwani, former director of the Central Soil and Water Conservation Research and Training Institute, Dehra Dun, India, for helpful discussions in the planning of this book.

Photographs and figures are by the author except as otherwise indicated.

Sir Charles Pereira FRS

Summary

Misuse of land on tropical watersheds in developing countries is increasing rapidly in association with accelerating population growth and the accompanying poverty. Destruction of natural vegetation without replacement by productive agriculture or forestry imposes severe penalties of soil erosion and sediment transport. Sedimentation destroys reservoir storage capacity and inhibits investment in power generation and irrigation.

Tropical meteorological events create seasonal water surpluses that produce floods in the lowlands. Flood abatement requires construction of storage structures in addition to good land use: only storage dams can both regulate flows and harvest the energy from them. Watershed management for stable conditions of vegetation, soils, and water is a critical requirement for investment in such flood protection.

Every corrective step to arrest watershed degradation improves the welfare of the inhabitants, both rural and urban:

- Restoration of a productive and protective tree cover on steep slopes can provide the fuelwood to forestall the impending crisis in the supply of domestic energy in developing countries.
- Planting of fodder and control of livestock can improve animal productivity and increase the supply of manure available for crops.

Technologies of soil conservation are well known but are durable only if the standard of farming is also improved. Productivity of crops and the stability of soils improve together.

In this book, the upper watershed technologies are described in plain language and successful examples are quoted. Where it has been possible to assess costs and benefits examples are quoted.

In the lowland reaches of tropical and subtropical river basins, the critical deficiencies are in the provision and maintenance of drainage. Salinity is the penalty for waterlogging: it denies to the third world the use of much of the great arid alluvial plains that could be irrigated by the Upper Ganges, Indus, Euphrates, Tigris, and Nile.

Distribution of water from irrigation canals between farms and within farms has been neglected as an aspect of management so that severe water losses and waterlogging occur widely. Successful solutions are described.

National, international, bilateral, and non-government agencies are making important contributions to the correction of watershed misuse, but are not coordinated either by country or by watershed. The national governments of tropical developing countries are not yet giving priority to rural area development in spite of growing food shortages. The primary cause of the watershed resource destruction in the tropical world has been rapid growth of population, which has overwhelmed government machinery for rural development and the management of natural resources. There is much confusion of emotion with moral judgment of this issue. No assumptions need be made, or should be made about the ultimate size of the population of any country. It is, however, an inescapable concern of common humanity that the arrival of extra millions of children should be delayed until arrangements can be made to feed them. In the meantime, irrevocable losses of soil, of water resources, and of reservoir sites reduce the agricultural and forestry production needed for the maintenance of growing populations. The race against time is being lost in Africa and is being narrowly held in Asia and South America.

Sir Charles Pereira FRS

Policy and Practice in the Management of Tropical Watersheds

Introduction

Before discussing policies of watershed management and the evidence for their consequences, it is appropriate to summarize the reasons why the fate of watersheds in the tropics* should indeed be a matter of concern to the whole international community. The tropics already hold some two-thirds of the human race and will produce four-fifths of the world's population increase in the next two decades. When the Club of Rome published *The limits to growth*(1) in 1972, their challenging analysis of world trends evoked much complacent comment. Extrapolations were said to make poor predictions, while the improving technologies of food production were confidently expected to meet the needs of the growing human population. In 1978 Wortman and Cummings published *To feed this world,* summarizing an impressive array of studies by the United Nations, the World Bank, regional development banks, the U.S. Department of Agriculture, the U.S. National Academy of Sciences, the International Food Policy Research Institute, and others.(8) These reports showed collectively that the world food problem is concentrated in the broad belt of tropical and subtropical developing countries. Unless major efforts are mounted on an unprecedented scale to improve land use and the productivity of agriculture and forestry in these countries, their poverty and scarcities of food, water, and fuel will increase to disaster levels.

At the same time, the low yields of tropical subsistence farming present a great opportunity for improvement. The protection and management of soil and water resources are the critical factors for success. In 1984 world grain production reached a

*The extent of areas under tropical meteorological conditions is shown in Figure 3.

record level in the high latitudes with an embarrassing 20 percent surplus for storage, yet the third successive year of drought has accelerated the time scale of agricultural decline for most of the African continent so that major food problems are already with us. The world total of grain production has little relevance to food shortages in the tropics; severe economic and logistic constraints prevent the continuous mass transfer of food from the higher latitudes from being a sustainable solution.

The 3-year drought that afflicted the African continent in the mid-1980s raised speculations about a change in world climate. There are only 200 years of rainfall data continuous enough for statistical analysis. The two major analytical centers for meteorological data are at Bracknell in Britain and at Boulder, Colorado, in the U.S.A. Scientists at both centers have studied the data and have found no evidence of general change in either the amount or the reliability of rainfall.

Mankind is, however, achieving an effect similar to world climatic change by the vast disproportion between the population growth rates in the high latitudes and in the tropics. UN estimates show that, in addition to the world's present 5 billion people, by the end of the century there will be at least another billion who will need to be fed. Some 80 percent of these will be born in developing countries dependent on subsistence agriculture: there they will be subjected to the inescapable hazards of drought and flood, which are characteristic of the Intertropical Convergence Zone (ITCZ). The overall effect of this differential population growth is to bring about a massive redistribution of humanity into climates less reliable for sustained agricultural production.

In spite of the encouraging progress toward stabilization of numbers by China and India, the population growth in more than half of the world's developing countries continues to outrun their food production so that *per capita* food supplies have fallen since 1975.(6) In Sub-Saharan Africa, the decrease has averaged 1.1 percent a year(7) (Fig. 1). The World Bank's *World development report 1984* emphasized the overriding influence of population growth in delaying the development of tropical countries.(5) The tragic instability of the present balance between population growth and food production in Africa has already been starkly

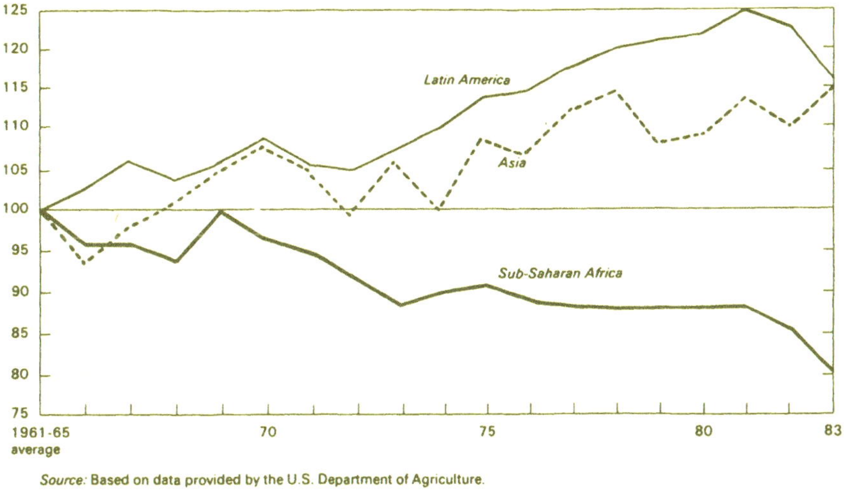

Figure 1. Index of per capita food production, 1961–1965 to 1983 (1961–1965 average = 100).

demonstrated in Ethiopia. Many further crises are inevitable if the present policies of both developing countries and aid donors continue. The symptoms of excess population pressure are the misuse of land and the rural poverty that ensues. When this occurs in critical upper areas of watersheds, the processes of hydrology transmit the damage to more productive areas downstream. Remedial policies to address both the reduction of population growth rates and the increase of food production are essential and urgent.

A critical aspect of subsistence agriculture under stress from excess of population is the progressive exhaustion of the soil's plant nutrients. Sparse crops increasingly expose soils to high intensities of radiation and of rainfall impact, accelerating the erosion of topsoil. As the overcrowded fertile areas are eroded, the additional population settles into poorer areas of drier or steeper soils in which the processes of land degradation are more rapid.

These effects, which now afflict some hundreds of millions of subsistence farmers, are not a recent development. They were rehearsed in the Mediterranean Basin more than 2000 years ago. Watershed degradation by forest destruction and overgrazing was eloquently described by Plato (*Criteas,* about 400 B.C.) who made accurate observations on the hydrological consequences of the deforestation of Attica. "The annual supply of rainfall was not then lost, as it is at present, through being allowed to flow over a denuded surface to the sea. It was received by the country in all its abundance, stored in impervious potter's earth, and so was able to discharge the drainage of the hills into the hollows in the form of springs or rivers with an abundant volume and a wide distribution." Lucius Columbella (*De re rustica,* about 50 A.D.) commented on the rapid decline of soil fertility after clearing of forest, "It is manifestly because of our own lack of energy that our cultivated lands yield us a less generous return; for we may reap greater harvest if the earth is quickened again by frequent, timely, and moderate manuring."(4)

For sustained high yields of food crops, a continuous input of plant nutrients is inescapable and some crop protection treatment is often necessary. Incurring such input costs is hazardous under erratic rainfall; inputs are most viable economically when water supplies can be assured. Irrigation is therefore of great importance under the erratic rainfall regimes of the ITCZ. The achievements in plant breeding, which have indeed merited in some regions the journalistic tag "Green Revolution," produced varieties highly responsive to fertilizers under irrigation and with ample sunlight. The combination of irrigation development and use of high yielding varieties has enabled both China and India to maintain food supplies in precarious but generally successful balance with their population growth. These two nations together farm some 70 percent of the world's irrigated lands.

In China, 1 billion people, about one-fifth of the world's population, depend on 107 million hectares of cultivated land, about half of which is irrigated for rice.(3) India since 1973 has annually developed over 1 million hectares of newly commanded and equipped irrigation projects. Massive financial support has come from the World Bank and the Asian Development Bank, which is currently investing in 41 irrigation projects throughout South and Southeast Asia. The ADB estimates that existing and potential deficiencies in rice production in the region will require investment of US $3.5 billion per year in irrigation until 1993.(2) Similarly, in South America, major irrigation developments are supported by international financial aid. The vast continent of

Africa has a limited area of well watered tropics, from which the Nile waters feed irrigation developments in the deserts of the North, but otherwise the resources of both surface water and groundwater are meager and scattered. Between the Sahara Desert and the Zambezi River lie 23 of the 36 poorest countries listed by FAO. In spite of the dearth of water resources in these countries, irrigation offers important prospects of increasing food supplies, even in the semiarid Sahel.

When the World Food Conference met in 1974 in the midst of a threatened food crisis, the document submitted by FAO indicated that a large part of the necessary increase in food production will only be possible through the rapid expansion and modernization of irrigated agriculture. A decade later the world food crisis has receded in global terms but has sharpened in Africa and Central America. Irrigation remains the key to much of the essential increases in food productivity.

Irrigation development usually occurs on valley floors and floodplains lying in the lower parts of the river systems. Some spectacular irrigation terraces have been developed on steep hillsides, but, in the major production areas, the hillsides collect the precipitation and deliver it to irrigate the lowlands. Dams and barrages control the flow and create reservoirs for seasonal storage. These reservoirs are totally vulnerable to flows of soil and rock debris and to the suspended sediment carried by the rivers. No economic solution to the engineering problem of removing accumulated mud and rock from reservoirs has yet been devised. Tropical countries have many filled and abandoned storage reservoirs as evidence of this danger: favorable geographic situations for dam building are scarce, so that the loss of the site may be even more serious than the loss of the investment.

Reduction of the rate of transport of soil and debris from hills to valleys is therefore a serious national problem for the many tropical countries whose future food supplies must rely on irrigation. On the hillsides, the loss of fertile topsoil has been a major factor in the impoverishment of subsistence farming communities. There are indeed many other reasons for the care and control of land use in the streamsource areas of river valleys, but the food situation in the tropics is the more urgent: the vulnerability of both irrigation and hillside farming to soil erosion by misuse of hill lands raises these issues to the level of international priority.

International recognition, accompanied by the organization and funding of several hundred remedial projects, has increased in the last decade, but so also has the scale of the problem. Rural rehabilitation schemes are treating many individual watersheds, but overall degradation of major streamsource areas continues unabated around them. A solution can be reached only by the governments administering these threatened river valleys. Awareness of the problem is the first stage in solving these essentially political issues. This study of the causes, with practical examples of successful remedies, is therefore set out in plain language for policy makers as well as for the wide variety of professional specialists who are involved. As a brief summary, drawing on my practical experience, it is offered to those who work in, plan, or fund development of tropical watersheds.

References

1. Meadows, D. H.; Meadows, D. L.; Randers, J.; and Behrens, W. W. 1972. *The limits to growth: A report for the Club of Rome Project on the Predicament of Mankind.* New York: Universe Books.
2. Oram, P., Zapata, J., Alibaruho, G., and Roy, S. 1979. *Investment and input requirements for accelerating food production in low-income countries by 1990.* Research Report No. 10. Washington D.C.: International Food Policy Research Institute.
3. Tang, A.M., and Stone, B. 1980. *Food production in the Peoples Republic of China.* Research Report No. 15. Washington D.C.: International Food Policy Research Institute.
4. Thirgood, J.V. 1981. *Man and the Mediterranean forest.* Toronto: Academic Press.
5. World Bank. 1984. *World development report 1984.* Washington, D.C.
6. World Bank. 1985. *World development report 1985.* Washington, D.C.
7. World Bank. 1986. *Poverty and hunger.* Washington, D.C.
8. Wortman, S., and Cummings, R.W., Jr. 1978. *To feed this world.* Baltimore: Johns Hopkins University Press.

1
Natural hazards in watershed development

The watershed: The natural basis for resource development

Land has been shaped by water throughout geological history, with small streams linking to form rivers or with gullies joining to form broad flood channels. The terms used in technical literature to describe the water-surplus areas from which streams flow may cause some confusion since usage varies. In the U.S.A., the whole area from which water drains into a river is described as the watershed. The high ground forming the boundary between adjacent watersheds is called a divide. In Britain, the area drained by a river is referred to as the catchment area or river basin; the high ground separating adjacent catchment areas is called the watershed. Because most of the earlier hydrological studies of land use were made in the U.S.A. and the American conventions are dominant in the scientific literature, they are used in the following discussions. The terms *watershed, catchment area,* or *drainage basin* will be used for the area contributing to a watercourse. The terms *divide* or *watershed boundary* will be used for the perimeter defining the area. The term *watercourse* will be used to include all natural channels along which water flows either perennially or intermittently. *Subwatersheds* are areas that generate separate streams within a larger watershed (Fig. 2).

In discussions of water resources, the term *river-basin development* usually refers to major river valleys. River-basin development studies characteristically include sites for major

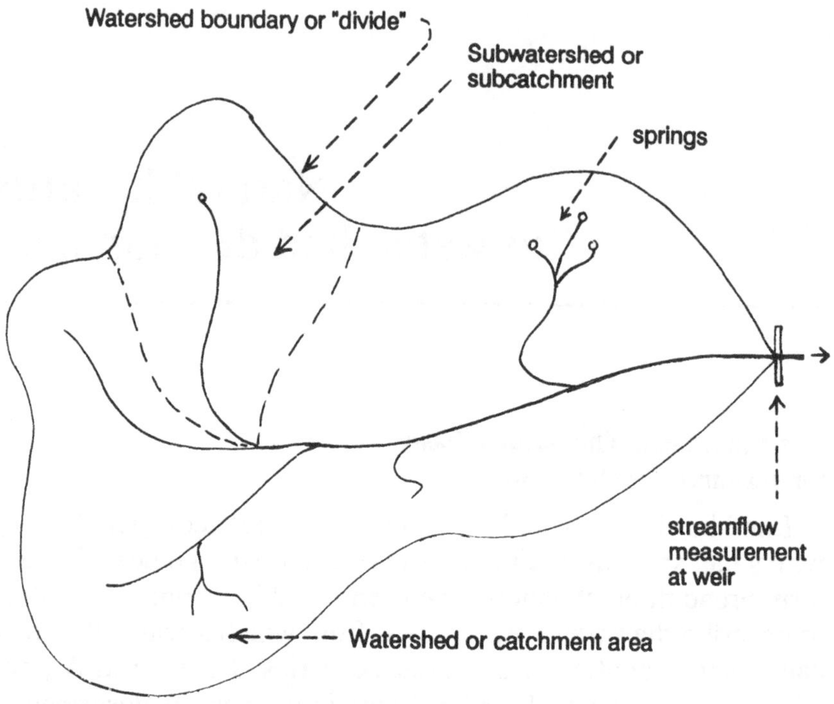

Figure 2. Definitions of watershed terms.

capital developments such as dams and reservoirs, for urban water supplies, hydropower, and irrigation. These require analysis of land-drainage systems and the exploration of underground water resources. This book, however, is concerned with the less-researched and often less-tractable land-use problems of rural areas involving the management of croplands, plantations, forests, and rangelands. These issues are not confined to the upstream areas because floods and droughts bring management problems to the whole of a river basin. The watershed is therefore the natural basis for management of land and water resources.

The major natural hazards of flood, drought, and sedimentation

Floods

The most critical aspects of upper watershed management, which directly affect the lives and property of downstream users, are the effects on floods, on water supplies, and on sediment transport. There is much confusion in the more popular writing on these matters, which is best explained at the outset. With very rare exceptions, such as the failure of major engineering structures, floods are not caused by human activities but by exceptional atmospheric events. Convergence of flows of moist air concentrate vast amounts of water vapor and cause its precipitation. Another natural cause is a sudden rise of air temperature over snow-clad mountains, as in the Himalayas, which results in a rapid thaw and the release of water at high elevations with high potential energy for the transport of soil and rock debris.

The evidence reviewed in the following chapters demonstrates that the misuse of land can greatly increase both the severity of floods and the damage caused. Soil and rock debris ripped from eroding slopes is deposited in reservoirs, over fertile farmland, on costly irrigation schemes, and even in city areas. Conversely, good use of land employing well-established techniques of forestry, agriculture, and civil engineering can substantially mitigate flood hazards, *but it cannot prevent floods.* Prevention is, of course, possible by building large dams where suitable dam sites exist and very large storage reservoirs can be created. The High Dam on the Nile at Aswan has eliminated the annual floods that had been recorded for more than 4000 years.

Floods do not always come from mountains or hills. Atmospheric disturbances that generate floods may occur over areas that are orders of magnitude greater than those of the watershed boundaries. Even in lands of low relief, however, the floods sweep along the drainage patterns of the watersheds. In 1979, the semidesert Indian border state of Rajasthan suffered a major flood at the beginning of the monsoon. This event was uniquely well documented because the new Central Arid Zone Institute at Jodhpur, developed by UNESCO and the Indian Council of Agricultural Research, had selected the watershed of the Luni River for hydrological study. Scientists had installed 240 rain gauges and several stream gauges.(3)

The onset of the monsoon was later than usual so that the development of a deep depression over the Rajasthan desert was to be welcomed. The vertical scale of the development was formidable, however. It extended upward 3 kilometers and joined with a center of convergence that was moving in on the western airstream from Pakistan. The combined cyclonic depression brought in strong winds laden with moisture from the warm Arabian Sea 800 kilometers to the southwest. For 5 days, this great vortex delivered rainfall over the 14,000 square kilometers of the upper watershed of the Luni River. Dry country where *annual* rainfall averages 300 to 500 millimeters received 700 to 925 millimeters in 5 days. Streams in the upper two-thirds of the watershed overflowed their banks to reach 5 meters above the normal levels and became 2 to 3 kilometers wide. In its lower reaches, the Luni River ran more that 25 kilometers wide. The death toll was 470 people and more than 10,000 head of livestock. Serious long-term agricultural damage was done by coarse sediments, which buried fields and irrigation systems under a layer 20 to 100 centimeters deep. These sediments lacked the clay or organic matter necessary to provide base exchange capacity and thus to hold crop nutrients, so that the fields became agriculturally useless.

Yet even under so massive an attack, good land-use practices were beneficial. Planting of trees such as kejhri (*Prosopis cineraria*), neem (*Azadirachta indica*), and bulbul (*Acacia nilotica*) proved to have been of great importance in reinforcing earthen banks of rivers and roads. Two popular fruit trees, *Ficus* and *Zizyphus mauritania,* proved to be very strong and saved many lives.(1)

Droughts

Drought incidence is also the result of events on a scale vastly greater than that of land-use problems. In tropical developing countries, the simulation of drought effects by the intensive misuse of land is a sadly familiar observation. The destruction of hillside forests, followed by unskillful cultivation and by severe overgrazing of hill slopes, can cause heavy rainfall to be shed as surface runoff instead of infiltrating to recharge the underground water reserves. As a result, the dry-weather flow of rivers may be diminished, and, in small watersheds, springs may fail and streams may dry up seasonally. Well levels fall and all the familiar signs of drought are manifest, yet the rain gauges show that the rainfall has not declined. Genuine droughts are caused by major perturbations of the atmosphere, which can affect very large areas, as did those that afflicted southern Africa in 1982, 1983, and 1984. As studies of global meteorology advance, heat storage in the surface waters of tropical oceans and the transmission of heat energy by ocean currents are increasingly believed to be implicated in drought phenomena. In energy terms, the atmosphere can be described as a vast heat engine driven by solar radiation with the oceans acting as a flywheel.

Sedimentation

Sediment transport is also a large-scale natural process. Although torrential flows of water and accelerated soil erosion are major penalties of land misuse afflicting many tropical developing countries, they are nevertheless the natural processes by which all lands were shaped. For example, the old rivers of the Tibetan plateau cut deep gorges through the rising new range of the Himalayas and spread vast quantities of sediments on the Indo-Gangetic Plains. The process continues today. In the lowlands, engineers build protection for the river banks, but in steep, rugged country the bank cutting and soil transport continues. Much of the watershed management technology discussed in the following chapters is aimed at reducing this erosion, but it can never be completely prevented. To this natural sedimentation humans have added vast increases from soil erosion by misuse of land in the upper watersheds.

The tropical aspects of the hydrological cycle: Advantages and limitations

The developing countries in which subsistence agriculture provides the dominant lifestyle lie mainly in the lower latitudes astride the equator. They extend somewhat further than the Tropics of Capricorn and Cancer; thus this discussion will concern the range of latitudes to approximately 30° north and south (Fig. 3).

The tropical climates within this range are dominated by high inputs of solar energy and the resulting rapid evaporation from warm surfaces of both sea and land. Together with the erratic patterns and high intensities of rainfall, these are inescapable characteristics of the Intertropical Convergence Zone (ITCZ),

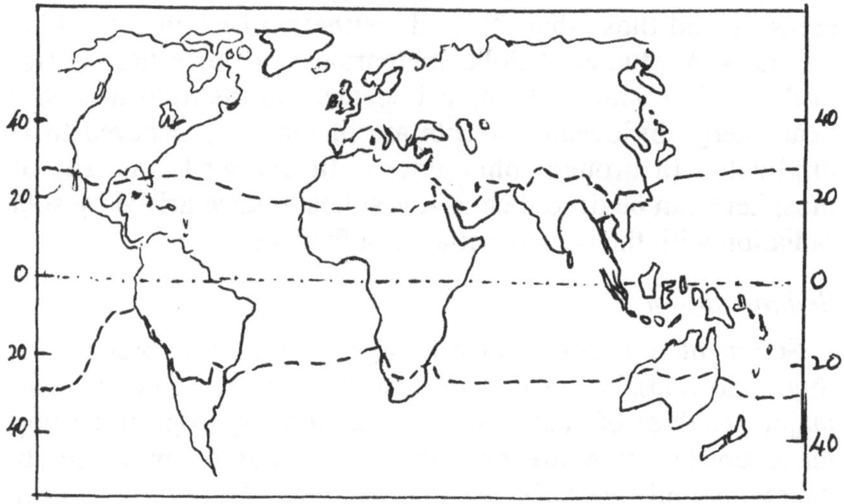

Figure 3. Tropical areas. The term *tropical* is not confined to the latitude limits of Capricorn and Cancer, but refers to the meteorological conditions that govern the broad irregular equatorial area between the northern and southern subtropical zones of the drylands. From these dry zones of high atmospheric pressure, the trade-winds converge to a fluctuating low-pressure trough of rising air that causes heavy precipitation. High levels of solar radiation dominate the climate in this zone, so that mean annual temperatures at low altitudes are characteristically above 20°C.

sometimes called the Intertropical Confluence between the trade wind systems north and south of the equator (Fig. 4). This irregular and discontinuous belt of low-pressure moist air fluctuates seasonally across the equator, following the sun. As the very large quantities of water vapor that have evaporated from warm tropical oceans condense into rainfall, they release latent heat energy, which increases the turbulence. Storms characteristically are heavy, with high intensities of rainfall. The ITCZ does not form a continuous trough or move in a regular or predictable pattern; it fills and reforms to produce a tropical chain of major disturbances, which cause highly variable rainfall regimes. Jackson gives a more detailed description of the ITCZ and its climatic effects.(2) Floods and droughts are thus inescapable features of the tropical latitudes within which a rising proportion of the world's population is born.

High temperatures afford great agricultural advantages: rapid plant growth and the possibility of year-round growing seasons.

Where rainfall is adequate or water can be provided by irrigation, farmers usually produce two crops a year, and three may be harvested under good management. Productivity per hectare per year can therefore be much higher than in temperate-zone agriculture.

The disadvantages are that evaporation rates are also higher, soil moisture reserves are more rapidly exhausted, and the high temperatures intensify the effects of drought. Because unpredictable drought is an endemic feature of many tropical climates, the hazards, especially for food-crop production, are frequently severe.

In tropical drylands (areas with less than about 600 millimeters annual rainfall), high temperatures and evaporation rates reduce the stability of vegetation cover, so that rangelands are more vulnerable to misuse. Overgrazing, for example, can actively extend areas of desert, as has occurred in the Sahel. Where bare soil is exposed, temperatures up to 70°C have been recorded. Organic matter is rapidly destroyed by oxidation under these conditions, thereby decreasing soil stability against erosion by wind, rain impact, and overland flow.

The wet tropics are also vulnerable to misuse of land because of the violence of tropical rainfall and the very large amounts of water to be disposed of. The characteristic vegetation is closed-

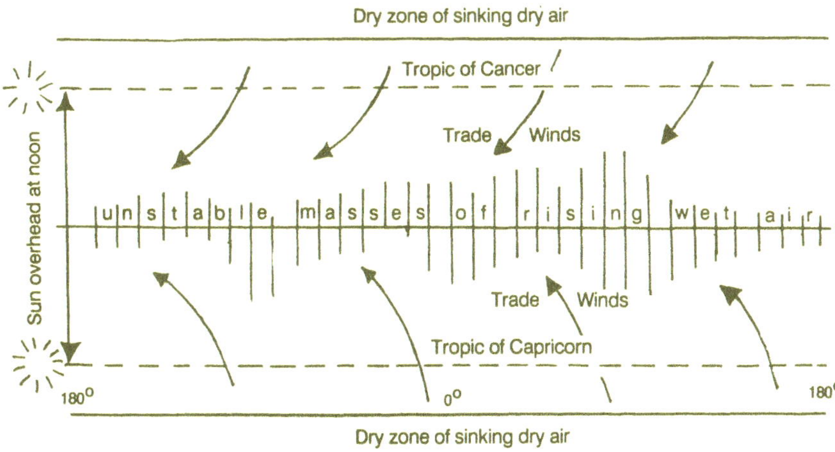

Figure 4. The Intertropical Convergence Zone—a zone of instability and turbulence characterized by droughts and floods.

canopy forest, which gives protection and stability. If the forest is removed without the introduction of adequate land-management practices, severe soil erosion is the sadly familiar consequence.

Water and energy cycles

Figure 5 illustrates the main features of the water cycle in the tropics. Although water transport by the atmosphere is a major feature of the cycle, the whole 4.5-kilometers depth of air holds, on the average, water vapor equivalent to only 25-millimeters depth of water. Some 97 percent of the world's water is contained by the seas, where an average of 3.5 percent (35,000 ppm) of dissolved salts renders it unfit for consumption by most land plants and animals. With energy supplied by solar radiation, water from the sea or land surface is evaporated, absorbing a vast store of energy as latent heat. Warmed by the surface waters of tropical seas, moist air rises into the cooler conditions above. Expansion causes further cooling, and the water vapor condenses into droplets as mist, fog, or clouds. Further upward movement causes precipitation as rain. Winds drive moist air against hills or mountain slopes, forcing it upward to cooler

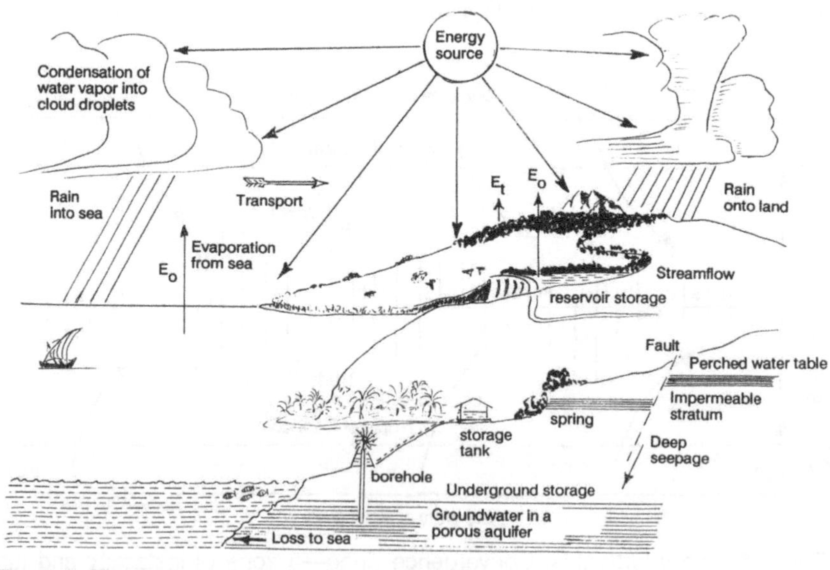

Figure 5. The hydrological cycle in the tropics.

levels. Rain therefore falls more frequently on hills or mountains than on nearby lowlands. In the higher altitudes of the tropics, hail is a serious agricultural problem (for example, in tea plantations). Snow fields and glaciers are features of tropical mountains above 5000 meters; the Himalayas, the Andes, and the high mountains of Africa (Kilimanjaro, Kenya, and the Ruwenzoris) all have permanent glaciers.

When rain falls on bare soil, it is either absorbed or shed as overland flow into drainage lines, which join to form streams. Rain falling on vegetation is partly intercepted, spreading to form a water film of large surface area that evaporates readily. Part of the water drains from the film as drips from the vegetation and part runs down stems or trunks to the soil. The water may be absorbed, or it may drain slowly through the litter layer, which acts as temporary or "detention" storage. Water infiltrating into the soil is in part stored within the root zone, to be later withdrawn by plant transpiration, or it moves slowly by gravity into deeper layers. Water penetrating through soil cracks, insect burrows, root holes, and so on moves more rapidly downward draining through permeable layers (sand, gravel, volcanic ash, or weathered rock) until it reaches an impervious stratum. This barrier may be solid rock or a clay layer such as buried lake or marine sediments. Above such a stratum, groundwater stored in porous layers, or aquifers, may drain slowly toward springs that maintain stream flow.

In the tropics, with seasonal alternation of heavy rains and hot, dry weather, these processes are of critical importance for maintaining surface water supplies. Surface conditions that permit acceptance and rapid infiltration of heavy rainfall are therefore essential to the recharge of aquifers and the continuation of spring flow in the dry seasons. Vegetation plays a vital part in the protection of the soil surface from the direct impact of rainfall and in the maintenance of rapid infiltration. Removal of such protection exposes the soil surface to damage by dispersal and capping, thus sealing the routes for water entry, smoothing the surface, and accelerating overland flow. Soil detachment and transport—the basic processes of soil erosion—then increase rapidly because the ability of water to carry soil rises in proportion to the fifth power of the velocity. Consequently, doubling the rate of flow increases soil transport capacity by 32 times. The

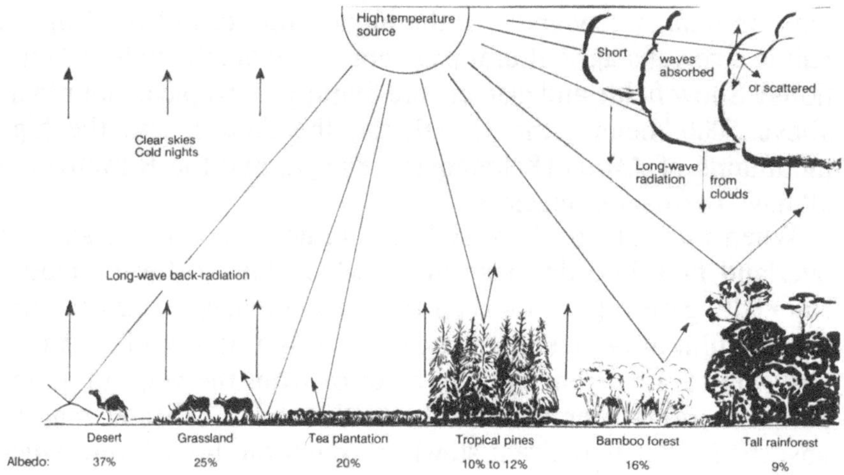

Figure 6. The energy cycle in the tropics.

problems of farming on hillsides in the tropics are thus an expression of the physical properties of soil and water.

Figure 6 summarizes the energy cycle that drives the water circulation. It illustrates why tropical crops sometimes lack optimum radiant energy in spite of the intensity of the tropical solar beam. Heavy cloud cover can intercept a high proportion of sunlight, as occurs on the island of Mauritius. There, mountains thrusting into the flows of warm wet air over the Indian Ocean lift and cool the airstream so that cloud cover restricts the yields of sugarcane, the country's main crop. The towering white cumulus clouds characteristic of the high plateau of southern Africa have an opposite effect: They reflect and scatter the shortwave radiation so that measurements at the crop surface show radiation intensities up to twice that of the direct solar beam.

References

1. Dhir, R.P. 1982. *July 1979 flash flood in the Luni: A case study.* Technical Bulletin no. 6. Jodhpur, India: Central Arid Zone Research Institute
2. Jackson, I.J. 1977. *Climate, water and agriculture in the tropics.* London: Longman.
3. Sharma, K.D., and Vangani, N.S. 1982. Flash flood of July 1979 in the Luni Basin. *Hydrological Sciences Journal (IAHS)* 27:285–398.

2
The impact of human occupation

Historical successes in hillside agriculture

Human occupation of a watershed need not result in degradation of soil and water resources, even under the extremes of tropical climates. Traditions for the safe cultivation of hillsides were evolved during early human history. The Ifugao terraces of the Philippines (Fig. 7), for example, have persisted for more than 2000 years. The discipline imposed by leveling the terraces to retain water has made hillside irrigation a highly stable form of land use even under acute pressure of population, as in Java (Fig. 8). As engineering skills have developed, tropical forests have been replaced by well-planned plantations of tea, coffee, cocoa, rubber, or oil palm that have flourished for more than a century. With high investment and competent design of roads, terraces, and drainage, together with skilled management, these plantations have maintained the stability of soil and water resources even in steep watersheds.

Modern capital development in tropical watersheds

At high levels of investment, as in large plantations of oil palm, rubber, or cocoa, the long-term stability of soils and water resources is required as the basis for future dividends. Such investment in tropical land development continues today with increasing participation by developing countries. A combination of skills in agriculture, forestry, civil engineering, economics, and marketing is basic to such development. Other inputs at the professional level, such as hydrogeological studies, remote sensing, and social and medical surveys are usually necessary. In

Figure 7. Ifugao terraces in Luzon, Philippines. These have permitted safe cultivation of hillsides for 2000 years despite annual rainfall exceeding 3500 mm.

tropical watersheds, plantations can safely replace closed tropical forest when good standards of management are maintained.

Erosion from mines and quarries

In contrast to plantation enterprises, the extractive mining and quarrying industries are major sources of watershed damage, particularly from open-cast or strip-mining such as the rather spectacular environmental damage from tinmining in Thailand. Even in Europe and the U.S.A., with long histories of industrial progress, adequate legislation to minimize ecological damage has been a late development. The strong political support needed to get such laws enforced has arisen even more recently. At mining sites, the problems of restoring vegetation and thus of achieving soil stability have proved to be severe. The subsoils and rock wastes stripped to expose ore bodies are often highly mineralized and are toxic to most plant species that volunteer from the surrounding area.

A particular problem of watershed management in mining areas is caused by sulfide ores, which become oxidized to sulfates when exposed to weathering. These compounds are leached into streams and cause acidity that destroys aquatic life. It is a problem in both temperate-zone mining areas and in the tropics. The main remedy so far found is to prevent the mine-wastes leachings from reaching the streams. Much research on restoration of strip-mined sites in the U.S.A. offers useful guidance to the study of such problems in the tropics.

Erosion hazards in road construction

The costs and difficulties of road building in steep hill country have ensured that most of the urban developments have remained in the milder topography of the lower reaches of rivers. The impact of ecological disturbance increases with distance

Figure 8. Intensive land use in Java. Holdings are commonly only one-quarter hectare, and terracing is meticulous.

A new oil palm plantation replacing closed tropical forest in Sabah provides complete cover. The plantation was developed jointly by the Commonwealth Development Corporation and a government-supported Malaysian company.

uphill because of the increased velocity of waterflow and the consequent increase in the transport of erosion debris. The Himalayas provide examples on a large scale. The Ganges loses more than nine-tenths of the height between source and sea in the first tenth of its course. As population expansion has pushed development further up into steeper country, road design and construction have become increasingly critical factors in tropical watershed management.

Even on forested hillsides, landslips and severe cutting of gullies are common consequences of inadequate engineering design of the outfalls from road culverts. Mountain roads act both as collecting surfaces and as hillside cutoff drains that intercept surface flows. When intense tropical rainstorms occur, roads concentrate water so that the culvert discharges have dangerously erosive energy. It is well within the routine technology of road engineering to provide for safe disposal of such flows by construction of channels protected by revetment with stone or concrete. These structures bring the water safely down the hillside to be discharged into stable natural drainage lines. Because such construction is expensive and can severely increase capital costs, road specifications in many developing countries limit the runoff protection to the minimum needed to protect the road foundation. Concentrated streams from culverts are thus launched down unprotected hillsides. Where sufficient labor is available, the cheapest method for stabilizing drainage routes is often the use of gabions. These are wire mesh boxes, woven on site and filled with broken rock or collected stones (see Chapter 8, Fig. 31).

On steep hillsides, even the best road designs have to expose some steep slopes of naked soil. The basic protection for these slopes is to dig interception drains above them, parallel to the road, to trap overland waterflow. The drains are led, preferably, to natural drainage lines that have cut down to rock and thus become stable. But the drains may have to be discharged down revetted channels to culverts under the road.

Protection of new road cuts and embankments by grasses is important even under temperate-zone rainfall, and in the tropics it can be critical. Broadcasting of seed with a small amount of nitrogen and phosphorus fertilizer in wet climates or planting of splits (rooted segments) of creeping grasses where long dry seasons occur is best done with tested local species, although two African grasses, *Eragrostis curvula* (African lovegrass) and *Paspalum notatum,* are widely used. A modern advance in the design of rural roads is the replacement of the deep narrow roadside drainage trenches, wherever space permits, by broad shallow grassed waterways, which are less susceptible to erosion. In tall forest, shading inhibits grass growth, but seeds of fast-growing tropical trees, of which Nepalese alder (*Alnus nepalensis*) is a good example, scattered over steep banks, result in protection by a network of roots.

The ability of four-wheel-drive vehicles to climb steep gradients on rough surfaces has led to much clearing of "jeep tracks" in tropical hill ranges. Unless these tracks are designed with skill, they are liable to become major sources of erosion. Although costs are reduced by steepening of gradients, it is a false economy to omit drainage protection. (The damage from poorly engineered temporary logging tracks is discussed in Chapter 4.)

The invasion of forested slopes

Small groups of forest-dwelling subsistence farmers have survived in tropical forests since prehistorical times by slash-and-burn methods of shifting cultivation. These practices consumed the accumulated fertility of soils and vegetation but affected so small a proportion of the upper watersheds that in the humid tropics the damage to soil and water resources was minimal. In more recent times, the numbers of shifting cultivators have grown; an FAO study estimated the totals to be of the order of 250 million people in 1974. In the drier open woodlands of the African savanna, the degradation is severe because they are burned annually. The ancient *chitemene* practice continues today in the *Brachystegia* woodlands of Zambia. Stems and branches are lopped from a wide area and burned to concentrate ash onto a small area on which millet is grown (Fig. 9).

Figure 9. The *chitemene* system in Tanzania has been practiced for centuries. Cultivators improve soil fertility before planting millet by cutting branches from a wide area and piling them between the tree stumps where they are burned and the ashes hoed into the soil.

TABLE 1
Annual deforestation in tropical regions

	Area (million hectares)		
	Closed forest	Open woodlands	Total
Tropical America	4.34	1.27	5.61
Tropical Africa	1.33	2.35	3.68
Tropical Asia	1.83	0.19	2.02
Total			11.31

Source: Food and Agriculture Organization and United Nations Environment Program. 1982. *Tropical forest resources*. Rome.

The rapid population growth in tropical developing countries in the last three decades has spilled very large numbers of subsistence cultivators into the remaining forests. As the soil on the lower slopes becomes exhausted, they increasingly invade slopes too steep for sustainable cultivation. A major survey of the status of tropical forests by FAO and the United Nations Environment Program (UNEP) (6) showed overall annual felling totaling 11.3 million hectares, which is projected to continue (Table 1). In peninsular Malaysia, for instance, of some 8 million hectares of high forest, most of the 3 million hectares on level terrain is scheduled to be cleared for agriculture; 4 million of the remaining 5 million hectares will be opened to logging. Here the cutting of vehicle tracks and the construction of temporary bridges open invasion routes to firewood contractors, illicit graziers, and uncontrolled slash-and-burn cultivators. Thus the degree of recovery after logging is highly uncertain. Logging on a sustained yield basis has a very poor record in humid tropical forests.(7)

An example of important international consequences from deforestation and subsequent soil erosion by misuse of watersheds is the Panama Canal. The watersheds that supply the reservoirs for the canal were under natural forest when the canal was opened in 1914 and remained stable for more than a half century. But uncontrolled settlement has destroyed the forest in recent years, and soil erosion has silted the reservoirs. The lessened infiltration has reduced the springflow that supplied the reservoirs that operate the great set of locks. The combined effect has been so serious that during a drought in 1977 the canal had to be closed to large vessels because water levels were too low. Technical assessment is that such closures may occur frequently in the future unless a forest cover can be restored on the critical watersheds.(13)

The most rapid losses of tropical forest are occurring in Paraguay, Costa Rica, Haiti and El Salvador in the Americas; in Nigeria and the Ivory Coast in Africa; and in Nepal, Thailand, and Sri Lanka in Asia.(11) The economic pressure of expanding populations to occupy land in upper watersheds will inevitably continue. All forested land suitable for sustained production of food crops will have to be cleared because the demographic implications of the present age structures of tropical populations are that massive increases will occur before the numbers eventually stabilize.(14)

An immediate result, already developed as a serious problem for many tropical populations, is the acute shortage of fuelwood (which is discussed in Chapter 4). Trees can be replanted while the soil remains, and even in the absence of trees, volunteer grasses and weeds rapidly provide a soil cover under tropical growing conditions. Competitive communal grazing in Africa, Asia, and tropical America has stripped and trampled the soil, exposing it to accelerated erosion by intensive tropical rainfall. This widespread degradation is most acute on the steepest slopes, so that the upper watersheds, remote from administrative control, have been particularly vulnerable.

In the middle mountains of Nepal, some 5 million farmers have so misused the steep slopes that their ability to feed their families has already become precarious, while their numbers continue to increase. In India, an estimated 13 million hectares of communal lands are now reduced to the condition of barren wasteland. India has 20 million hectares of gazetted forestland administered by federal or state forest services. One-third of this forest has been so severely damaged that active soil-conservation work is now required to replace the protection formerly provided by the trees. In the vast area of densely populated Indian agricultural lands, government surveys estimate that 58 percent are subject to severe erosion.(12)

Overcrowded tropical lands are not all thus misused. Indo-

Overgrazing of steep lands has initiated gullies that have cut through the crop terraces.

nesia offers some remarkable contrasts. Java has a population of 91 million of whom 70 percent live by farming at an average density of 600 people to 1 square kilometer. Average family holdings are 0.6 hectare on the rainfed lands or 0.26 hectare in irrigated areas. In spite of this intensive population pressure, the steep upper slopes of the many volcanos remain under protective forest cover, with 21 percent of the land area forested. Where

Figure 10. Tropical tree garden in Java. Sheep and goats are stall-fed in raised thatched pens or are allowed to graze on tethers.

these protective forests have been damaged by illicit felling, they are being replanted under the *taungya* system, which organizes peasant farmers to plant trees and grow crops between them until the tree canopies close.

Typically, the family holdings of less than 1 hectare have every square meter in production in multistoried tree gardens. Coconuts, bananas, mangoes, and jackfruit are underplanted with the ubiquitous cassava, sweet potatoes, and yams. Trees also provide cash crops of cloves, robusta coffee, cocoa, citrus, and pepper (Fig. 10). Livestock are stall-fed or tethered. There is no free-range grazing, a fortunate tradition arising from sheer necessity because the land is held individually. The irrigated areas are intensively developed and highly productive. There are serious problems of land misuse and soil erosion on the poorer and steeper lands. These problems are associated with the practice of sharecropping in which absentee landlords are more concerned with payments from their tenants than with care of soils. Soil washed from hillsides overcultivated for cassava has already damaged some intensively developed irrigation areas. This trend is a serious concern for the government, which supports the highly productive irrigation areas with subsidized fertilizers and insecticides and is cooperating with overseas aid teams to restore soil stability to the eroded upland areas.

In Sabah, the oil palm is one of the many tree crops grown by smallholders in tree gardens; the fruit bunches are collected from the roadside for processing by government-owned factories. Following the practice of the large estates, the smallholders are increasingly using fertilizers.

Similar tree gardens are a feature of the intensively settled slopes of the Kandy area of Sri Lanka on formerly forested land. A rainfall of about 1500 millimeters and access to markets appear to be the main determinants of tree-garden economies. They represent a productive and sustainable adaptation of natural forest ecology to support a high density of human population.

The exhaustion of soils

Early historical records of farming methods show that the maintenance of soil fertility by the use of animal manures to replace the nutrients extracted by crops has been practiced for at least 3000 years in the Mediterranean region and probably for 4000 years in China. Virgil's *Georgics* contain specific instructions for application of manure. Yet failure to replace crop nutrients is a principal cause of the land degradation that increasingly threatens tropical watershed management.

Despite their luxurious natural vegetation, many tropical soils have low levels of plant nutrients. Tropical forests characteristically work on a closed cycle between the biomass and the topsoil, as nutrients are taken up through the dense mat of surface roots and returned by leaf fall, as shown by studies in Belgian Congo, Ghana, and Trinidad.(1, 8, 9) When the trees are felled and burned, the mineral nutrients in the ashes and the topsoil are exposed to heavy rainfall, with severe losses by leaching and by soil erosion. After only two or three crop seasons, the yields decline, weeds accumulate, and farmers abandon the area to forest regrowth.(10)

When population growth forces more or less continual cultivation, crop growth becomes so poor that the sparse cover cannot protect the soil surface from sun and rain. Soil organic matter is rapidly oxidized away, and with it is lost the critical ability of the soil crumbs to maintain their integrity under the battering of tropical rainstorms. The soil aggregates are dispersed to paste that seals the surface so that runoff replaces infiltration. Annual loosening by cultivation and transport by sheet erosion rapidly remove the residual fertility from the forest soils. The soil depletion may reach a stage at which not even weeds will grow, although a few deep-rooted trees may survive. The effect on the hydrological behavior of the watershed is a progressive increase in the proportion of rainfall that is shed as runoff, with acceleration of the rates of flow and transport of soil. This damaging process has already affected a high proportion of the developing countries of the tropics and subtropics. Current rates at which cropland is being thus destroyed are not well established, but FAO estimates that 10 million hectares are lost each year.

Semiarid watersheds under subsistence agriculture are particularly vulnerable because livestock play a dominant part in the farming systems; traditional grazing methods have been overwhelmed by the rapid expansion of both human and animal populations. Sustained productivity requires the limitation of animal numbers to the carrying capacity of the land. This is a difficult problem (discussed in Chapter 7), but the social organizations in more than 100 developing countries have not yet found alternatives to competitive overgrazing of common land. The

(Left) Soil exhaustion from shifting cultivation (granite sands in Zimbabwe); (right) response to N + P fertilizer.

most serious damage is in the Sahel and many of the African countries to the south, in the Middle East, and in northwest Asia. The process is accelerating and more than 20 percent of the earth's surface, occupied by some 80 million people, is under attack.(14) The degradation process is often referred to as desertification, but in true deserts, the soil may be quite fertile and may respond well to irrigation. The eroded, barren lands created by misuse have often lost too much soil for rapid recovery.

Remote sensing by satellite imagery, aerial photography, and side-looking radar is providing increasing evidence of the scale and gravity of the impact of human populations on their environment. The complexities of ecology and of human land use in the rural areas of the tropics prevent any useful attempt at a quantitative assessment of soil erosion in the tropical world. The literature abounds with subjective estimates. Measurements by engineers of the soil transported by rivers or trapped in reservoirs give the most effective quantitative indications, but even these are difficult to interpret. The maximum transport of soil occurs in the transient peaks of stormflow, which are often missed by routine sampling, whereas unknown accumulations of bedload (coarse sand, gravel, and rocks) and of riverbank deposits obscure the meaning of quantities of soil in suspension. In the literature of soil erosion losses around the world, the most comprehensive recent review, *The state of the world 1984*,(2) estimates that the accelerated erosion loss is in excess of the rates of natural geological erosion, by 25.7 billion tons a year. The FAO publications *Agriculture towards 2000*(5) and Soils Bulletin 33(4) provide a well-informed perspective, and Brown and Wolf's paper, *Soil Erosion*,(3) gives a brief and clear summary. To those who study the evidence, the threat to the human race of the misuse of land ranks with the more dramatic threat of modern warfare. Both are avoidable, but neither will be avoided unless the challenge to achieve better solutions is taken up seriously and urgently.

References

1. Bartholomew, W.V.; Meyer, I.; and Laudelot, H. 1953. *Mineral nutrient immobilization under forest and grass fallow in the*

Yangambi (Belgian Congo) region. Serie Scientifique no. 57. Leopoldville, Belgian Congo: Publications of the Institute National d'Étude Agronomique (in French).
2. Brown, L.R.; Chandler, W.; Flavin, F.; Postel, S.; Starkel, L.; and Wolf, E. 1984. *The state of the world 1984*. New York: Norton.
3. Brown, L.R., and Wolf, E.C. 1984. *Soil erosion: Quiet crisis in the world economy*. Worldwatch Paper 60. Washington D.C.: Worldwatch Institute.
4. Food and Agriculture Organization. 1977. *Soil and water conservation and management in the developing countries*. Soils Bulletin no. 33. Rome.
5. Food and Agriculture Organization. 1979. *Agriculture towards 2000*. FAO Conference, 20th Session, 7924. Rome.
6. Food and Agriculture Organization and United Nations Environment Program. 1982. *Tropical forest resources*. Forestry Paper no. 30. Rome: FAO.
7. Fox, J.E.D. 1976. Environmental constraints on the possibility of natural regeneration after logging in tropical moist forest. In vol. 1 *Proceeding of the sixteenth IUFRO world congress, Oslo,* 512–536. Munich: International Union of Forest Research Organisations.
8. Greenland, D.J., and Kowal, J.M.L. 1960. Nutrient content of a moist tropical forest of Ghana. *Plant and Soil* 12:154–174.
9. Hardy, F. 1936. Some aspects of tropical soils. In vol. 2 *Transactions of the third international congress of soil science,* 150–163. Amsterdam: International Society of Soil Science and Royal Tropical Institute.
10. Nye, P.H., and Greenland, D.J.. 1960. *The soil under shifting cultivation*. Technical Communication no. 51. Harpenden, U.K.: Commonwealth Agricultural Bureau.
11. Office of Technology Assessment. 1984. *Technologies to sustain tropical forest resources*. Washington D.C.: Congress of the United States.
12. Randhawa, N.S. 1980. Watershed development in India: An Overview. In *Proceedings of the national symposium on soil conservation and water management in the 1980's*. Dehra Dun: Indian Association of Soil and Water Conservation.
13. U.S. Inter-Agency Task Force. 1980. *The world's tropical forests: A policy strategy and program for the United States*. Department of State Publication no. 9117. Washington D.C.
14. World Bank. 1984. *World development report 1984*. Washington, D.C.

3
The scientific basis for management of tropical watersheds

Effective planning of land use on watersheds must be based on quantitative information on the physical resources of climate, topography, geology, soils, vegetation, and water resources. If watersheds are inhabited, then further information is required on the numbers and distribution of people and animals, communications, economic and subsistence landholdings, and farming systems. Local history and political structures need to be understood in order to organize cooperation to improve land-use practices. The physical measurements alone are formidable enough in the rugged, forested hill country, remote from roads and towns, in which most major tropical rivers arise. Even more difficult are the social and economic complexities of heavily settled upland watersheds. Many of these produce, in spite of mild topography, a series of muddy torrents and meager, polluted dry-season flows, which are unacceptable to populations of the lowlands.

Reconnaissance surveys

Formal organization and funding of field studies by multidisciplinary reconnaissance teams to assemble this information have become accepted as necessary components of investment in watershed development. It is equally necessary to build up such information to improve administration of land use in watersheds, even in the absence of funds for economic development.

Topographic mapping from aerial photographs by stereographic plotting has already been achieved throughout much of the tropics; experts and facilities are available on contract to complete the task. Where cloud cover is persistent, as in the forests of Papua–New Guinea, side-scanning radar can provide excellent accuracy and continuity. Geological surveys, at least at reconnaissance level, have also been completed for most major tropical areas in which watershed management is likely to be organized. A hydrogeologist is, however, usually a necessary member of the team undertaking reconnaissance studies for watershed management, in addition to a surface-water hydrologist.

The most useful clues to the potential for agricultural or forestry production are provided by the undisturbed natural vegetation. Ideally, on undamaged watersheds, ecological transects relating soils to natural vegetation give the most positive indications of suitability for other land uses. Where most of the natural vegetation has succumbed to fire, ax, or livestock, it is worthwhile to make considerable efforts to reach and to study remnants that have survived on sites difficult of access. Sacred groves surrounding shrines protected by religious traditions offer useful evidence of the potential for tree growth in a denuded landscape.

Field measurements needed for good management

The basic components of the hydrological cycle should be the concern of anyone responsible for watershed management at any level—an interdepartmental committee for a river valley, an administrator of a hill district, or a lone forester stationed in a streamsource area. All should be familiar with the annual summary:

$$P = Q + E + \Delta S + \Delta G$$

where P = precipitation, Q = streamflow, E = evaporation, ΔS = change in soil moisture storage, and, ΔG = change in groundwater storage. These terms summarize complex natural processes, so that measuring them on a watershed scale requires substantial effort. They lie in the technical fields of meteorology,

Trees protected by a shrine show the potential for tree growth in this denuded landscape of the Ethiopian highlands.

hydrology, civil engineering, geology, and soil science. Watershed managers cannot master all of the disciplines, but they must know when to seek expert advice. The following pages offer help in understanding that advice.

When a seasonal cycle of rainfall prevails, the *water-year* is calculated from the average date of minimum streamflow. Both soil moisture and groundwater storage then show minimum annual variation. A crude approximation thus gives $P–Q$ as the evaporation or water use from a valley. This approximation assumes that storage changes are negligible and that there is no significant leakage from the bottom of the valley or under the measuring weir. For investigational work, much closer study is required.

Rainfall

Inquiries at urban offices about rainfall in remote forested hill country all too often elicit the response that "no information is available." This answer should not be accepted without further investigation. Outposts in the hills maintained by foresters, police, army, or customs staff should be visited. Sometimes individuals set up rain gauges and read them for years without anyone in town being interested enough to collect the data.

Rainfall amount and seasonal pattern are such critically important pieces of information that measurements should begin at the earliest stage in programs of watershed management or development. Yet this rarely happens. Too often, it is only after plans, negotiations, surveys, funding delays, and road building that measuring stations are set up as the last stage in a project that may have been in preparation for several years. Serious mistakes in planning may well have been made in the meantime from wrong assumptions about the amount and distribution of rainfall.

To monitor the environment, well-educated field staff and sophisticated equipment may eventually be employed, but in the tropics the former are usually townspeople who fear and dislike lonely situations in forested hills. I have experienced greater success by recruiting people from the fringes of the forest and training them to carry out simple routines. Effective checks on consistency can be arranged, such as the use of storage gauges read by the supervisor to validate daily readings by the observer. The storage gauge, with a little oil floated on the reservoir, will read only slightly less than the total of the daily readings. If it should indicate more, it suggests that some daily records have been missed.

Rain gauges are deceptively simple devices. They take very small samples of immense populations of falling raindrops. The catch is strongly affected by wind speed. Because this is lowest at ground level, due to the friction drag on the airflow, gauges are sited as low as possible, without collecting splashes from the ground. The World Meteorological Organization (WMO) recommends a standard height of 30 centimeters for gauges set in short grass, but for less standard conditions in watershed work, the U.S. Department of Agriculture recommends 75 centimeters.(20) Where strong winds are frequent, as at coastal or mountain sites, the gauges should be shielded by a cylindrical funnel or by a series of hinged metal strips (for example, the Altar Shield recommended by WMO). Shielded gauges may catch up to 5 percent more rain than unshielded instruments on a windy site. They are essential for snow measurement, but not necessary for most inland tropical conditions.

The diameter of the gauge makes surprisingly little difference for accuracy. At the turn of the century, studies by Mills showed that rainfall collected by gauges ranging from 7.5 to 60 centimeters in diameter varied only by 1 to 2 percent.(7) Huff compared gauges of small diameter with the U.S. National Weather Service standard 20-centimeter gauge and found both the 7.5-centimeter circular gauge and the 5.8- × 6.4-centimeter plastic wedge-shaped gauge to be satisfactory.(5) However, the plastic gauge loses water by evaporation too rapidly for watershed monitoring in tropical conditions.

On forested sites, small clearings are necessary for routine monitoring. Conventional meteorology prescribes that the distance of the trees from the rain gauge should be at least twice the tree height. This guideline was based on early work on the Great Plains of the U.S.A. For continental tropical sites, wind speeds are low (an annual average of about 7 kph is characteristic). For these conditions, experiments by McCulloch in East Africa have shown that a smaller clearing, giving a 1:1 ratio, is adequate(6) (Fig. 11).

The numbers of rain gauges required depends largely on whether the results are to be used for research, such as water-balance studies, or for general purposes of watershed management. Tropical rainfall is predominantly of a convectional type that produces characteristic thunderstorms with high intensities over limited areas. A network of gauges is therefore necessary so that the well-known Theisson Polygon method of calculation can be used to estimate the input to a watershed. This method requires some gauges to be near the perimeter, either inside or outside of the catchment area. Where the vegetation growth indicates that a rainfall gradient exists—for example, as rainfall increases with altitude in steep valleys—the differences due to gradient must be statistically eliminated from those due to random error. This can be done by dividing the watershed into broad horizontal strata with at least two gauges in each. In experiments in East Africa, dividing a steep valley into three strata and siting two rain gauges in each reduced the standard error of the mean

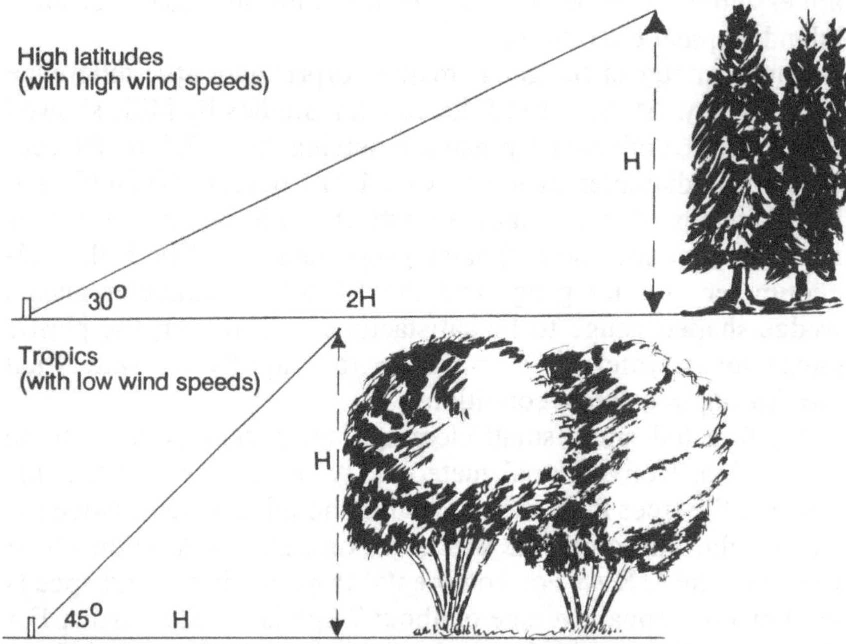

Figure 11. In choosing or clearing a site for a rain gauge, obstacles must be far enough away to avoid windshelter effects. In high latitudes, the 30° limit is used. Because wind speeds are lower at inland sites in the tropics, a 45° limiting angle is adequate.

catch for the valley by half. Later, when the forest was cleared, another 15 rain gauges were set up. They gave the same average as the original six. When in doubt as to the numbers required, a good principle is to put in extra gauges and observe whether they alter the mean catch.

Aspects of the hillsides may also be important because slopes are affected by prevailing winds. Strong contrasts are observed along the flanks of the Himalayas. Slopes should therefore be gauged separately where possible—that is, gauges should sample representative areas of the watershed. Within these areas, the sites should be chosen at random, but formal randomization is less important in practice than is reliability of observation. Access from a vehicle track or good footpath is of great importance for long-term reliability. The human frailties of the observer, however well-trained he or she may be, are the most critical factors in the reliability of data. It is unwise to site rain gauges that are very difficult to reach.

Watershed managers will find that the technical literature describes many devices for improving the accuracy of rainfall measurement. Gauges mounted parallel to the slope, set in shallow pits, or surrounded by grids of "eggbox" cells all have their places in research programs, but they are not needed for management purposes. Where soil erosion is a dominant concern, information on the intensity of rainfall is important. Several well-known "tilting-bucket" instruments recording on clock-driven charts are widely used in the tropics and require only a modest degree of training for observers. But the time necessary to interpret the charts may result in processing backlogs unless there is strong supervision. Modern instruments especially designed for tropical use employ solid-state electronic circuitry to record the tilting bucket oscillations on magnetic tape. Cassettes are easy to change. These are more reliable than clock-driven instruments, but they need compiling and computing equipment in the office, which in turn requires a higher standard of staff training. The urgent need to reduce soil erosion as a basic step for expanding food production in the tropics is now receiving world recognition. As a result, data collection for rainfall intensity will undoubtedly become more widespread.

Streamflow

Engineers in the water supply, power generation, irrigation, or water transportation departments are concerned with supply, storage, and maintenance of minimum flows. Their routine measurements of streamflow may be useful sources of data for watershed management. Many such measurements have been developed on "representative basins" in tropical countries by the UNESCO International Hydrological Decade and the subsequent International Hydrological Programme. Such studies are, however, rarely equipped to measure water balances or the hydrological effects of land-use policies.

Where water is scarce, watershed managers need to know how contrasting systems of land use, such as cropping, grazing, plantation development, or natural forest, affect streamflow. To gather this information, measuring devices must be sited where an impervious rock formation or clay deposit brings the whole output of the watershed into the stream. Estimates of changes in groundwater storage and soil moisture are needed to reduce the uncertainties of the water year, but greater errors may arise from deep seepage through the bedrock of the valley. Although this seepage cannot be directly measured, an effective check is given by an independent assessment of the total evaporation loss by means of the energy balance.

A more obvious source of error, which can usually, but not always, be avoided by competent hydrologists, is leakage into shallow aquifers that underpass the measuring device and rejoin the stream further on. The essential precaution is to take measurements at intervals along the stream and to plot a profile that will show sudden gains or losses. I have had the sad experience of visiting a well-constructed and expensively equipped weir high in a forested watershed where this danger was overlooked. Inspection of the catchment area suggested that the flow was far too small. Exploring some 200 meters downstream in dense thicket revealed a spring that doubled the flow.

The costs and accuracy of devices for measuring streamflow vary widely. For minimum capital outlay, daily readings of a staff gauge, which is a fixed or hand-held depth scale, may be taken at a point where the streambed and banks are stable and turbulence is minimal. Economical as this may seem, it requires many visits at different stages to measure streamflow with current meters and thus to build up a rating curve relating depth to flow. The high points on this curve occur in major storms in the rainy season when steep tracks are difficult or impassable. If such stormflows erode the streambed or deposit debris, a new calibration or even a new site may be needed. For large rivers with stable beds and banks, the rated section is the usual method of measurement: Depth of water is registered by a float in a stilling well, whose rise and fall are recorded on a clock-driven chart, a punched tape, or a magnetic tape.

Long experience, particularly in the U.S.A., in the gauging of upland watersheds has produced well-known designs for the installation of sharp-crested weirs. These weirs have a notch that is V-shaped, rectangular, or a combination of shapes, cut in a heavy metal plate and mounted in concrete or masonry (Fig. 12).

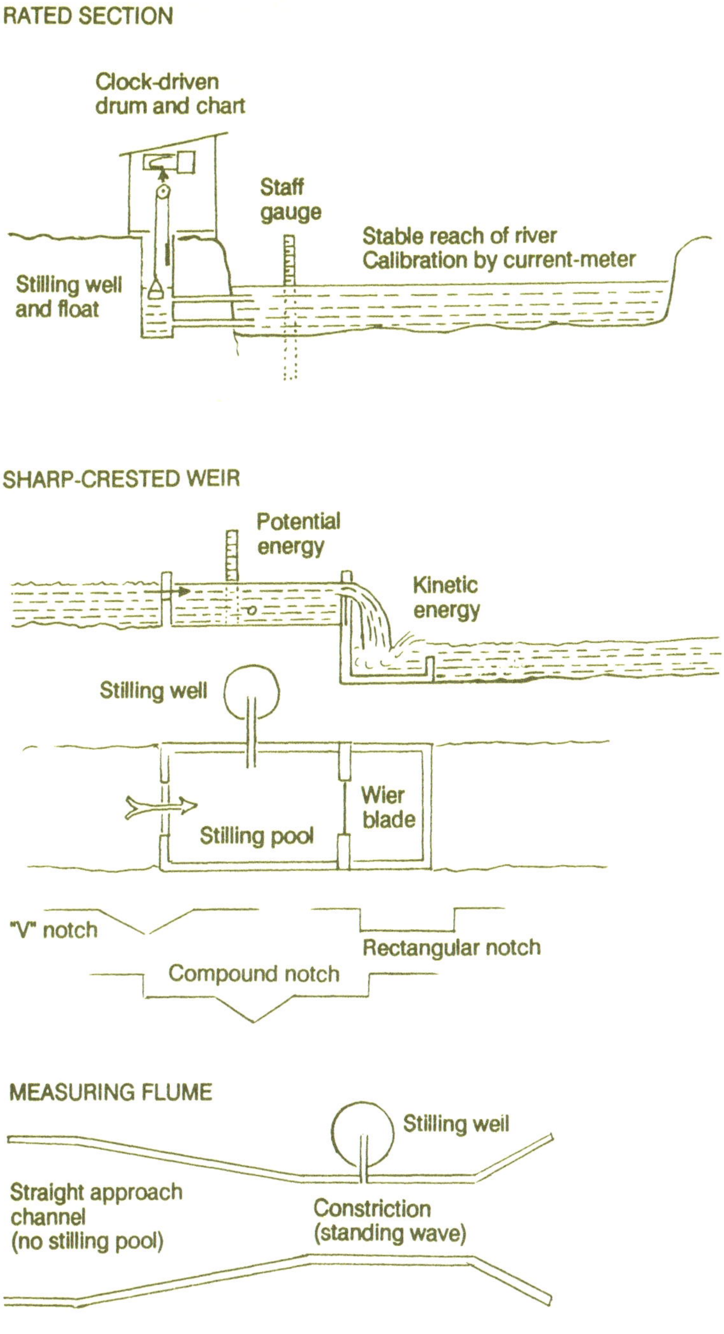

Figure 12. Some methods of measuring streamflow. In the tropics, most routine hydrological records are taken from rated sections. Many forms of weir and standing-wave flumes can be found in engineering literature.

The waterflow must first be arrested by a stilling pool, which may have baffles to reduce velocity. The flow is then related by empirical formulas to the difference in height between the surface of the pool and the bottom of the notch. The formulas are based on the transition from the potential energy of the water in the pool to the kinetic energy of the jet issuing from the notch. It is essential to remember that these empirical formulas are based on prescribed conditions of flow, which are achieved by the design and dimensions of the structure. These specifications are not easily or cheaply attained in steep, rugged streambeds, but variations can cause wide errors so that the data have no quantitative meaning.

Measuring flumes are used where rock debris, soil, or trashflows occur or where tropical rainfall causes flash flows from small watersheds. They are variously referred to as standing-wave, critical-depth, or hydraulic-jump devices. These devices are open channels that require a straight-channel approach but do not need a stilling pool. They are shaped to give a constriction, either from the sides or by a hump in the bottom of the channel. The flume holds water back on the upstream side, while the velocity of flow increases through the constriction until a critical depth is reached and a standing wave occurs below it. Energy principles define the relationship of the flow to the depth, so that calibration is not needed. There are several useful designs. The U.S. Soil Conservation Service developed the widely used "H" and "HL" flumes. Channels of trapezoidal section are also convenient for construction.

All measuring flumes need accurate construction. The smaller sizes are prefabricated in wood or metal or as moldings of acrylic resin reinforced with fiberglass, a technique developed for boats. The molded trough is light in weight and is bolted to angle-iron supports set in concrete. A strong advantage of prefabricated flumes is that they are easy to install, whereas sharp-crested weirs require more skilled artisans on site.

Water levels are usually recorded by pen onto clock-driven charts; electronic equipment, which uses magnetic tape or punched-paper tape, is also available. Sophisticated electronic recording devices however require trained observers and supervision by professional staff who understand the equipment.

Periodic visual and manual checks of water levels should be noted in the inspection log. This information permits valuable corrections when calculating the results. I know of an instance in which a set of new instruments was provided by a donor, but no checks were made, and the punched-paper tapes were in due course found to have recorded nonsense throughout a whole season's records of an expensive investigation.

Sediment flow

Soil particles detached by the impact of rainfall or eroded by flowing water are carried along in the stream as suspended sediments. Heavier particles of coarse sand, gravel, or rock debris are rolled and bounced along the streambed as bedload. Both have important effects on the life expectancy of reservoirs and on the maintenance of irrigation systems. Sediment flows are critical indicators of the effectiveness of watershed management. Methods of measurement are, however, too difficult and time consuming to be undertaken readily at remote stations. The simplest routine method—taking daily samples of 1 liter or less of water—misses the occasions of massive soil transport by stormflows. Metal intake tubes can instead be used to fill bottles at preset heights from a rising storm flow (Fig. 13). The state-of-the-art solution employs automatic, electronically controlled intakes that are connected in turn by a rotating manifold to sample bottles from which the air has been removed with a hand pump. Samples are thus taken at preselected stages of the rising and falling stormflow peak. All of the samples must be filtered, oven dried, and weighed; as a result, the overall cost of such routine measurements is substantial. Yet the large investment in reservoirs and irrigation schemes usually provides ample justification for such monitoring programs. The trapping and measuring of bedload are even more difficult, and methods are improvised to fit local situations rather than being standardized.

Evaporation

Evaporation accounts for most of the difference between annual totals of rainfall and streamflow. The rest is due to storage changes in the watershed or to leakage from the streambed. Evaporation occurs directly from wet surfaces of soil or vegeta-

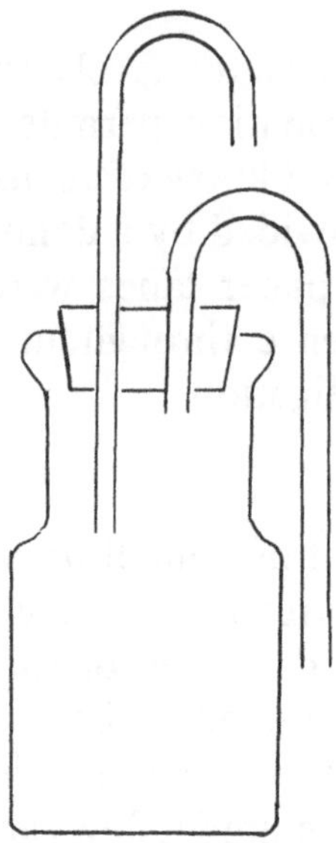

Figure 13. Sampling tubes for measuring sediment in stormflow. A set of samplers is mounted vertically through the expected range of water levels. This equipment gives only approximate values for suspended sediment.

tion and less directly through the stomata of transpiring plants, which control the process by opening or closing. The term *evapotranspiration* has sometimes been used to include both processes, but this is unnecessary. The vapor losses from a watershed are correctly ascribed to evaporation.

It is, however, necessary to distinguish between "potential" evaporation (E_O), which is the loss from an open water surface (a pond or lake rather than a beaker), and actual evaporation (E_T), which is the loss from the land surface of the watershed. Potential evaporation depends on the energy available to supply the latent heat for evaporation and on the ability of the atmosphere to accept and to remove the water vapor. Actual evaporation is governed by the availability of water in the soil and vegetation to meet the demand presented by the weather conditions.

Estimation of evaporation in the field is a rather complex procedure that is now understood in physical terms. It is accepted as a routine necessity for the operation of irrigation systems and storage reservoirs. Knowledge of the evaporation rates over a watershed is necessary for the development of water-resource policies.

A simple device that can give a rough estimate of losses from open water is the evaporation pan. The World Meteorological Organization has chosen the U.S. Weather Bureau Class A Pan as a world standard. It has a 1.22-meter diameter and is mounted on a grid of timber supports so that the base is well ventilated and may be inspected for leaks (Fig. 14). Because the pan is heated by the sun from the sides as well as from the surface, the rate of evaporation is about 40 percent higher than the rate from an equivalent area of a lake surface. The readings are therefore reduced by a "pan factor" of 0.7, based on the summer values measured by the U.S. Geological Survey on Lake Hefner (Oklahoma).(21) In the tropics, protection from drinking by birds and animals and from the mass incursion of moths may be achieved by covering the pan with a thin-wire screen of 1-centimeter mesh. To avoid abrupt changes of reflection when the galvanized iron floor is periodically cleaned of mud, the interior should be coated with bitumen, which also reduces corrosion. The small changes in readings caused by these two modifications are opposite and approximately equal.(13) Although they are crude devices, evaporation pans have often provided engineers in the tropics with their only quantitative clues about evaporation when designing reservoirs and irrigation systems.

Advances in the application of physics have now made better

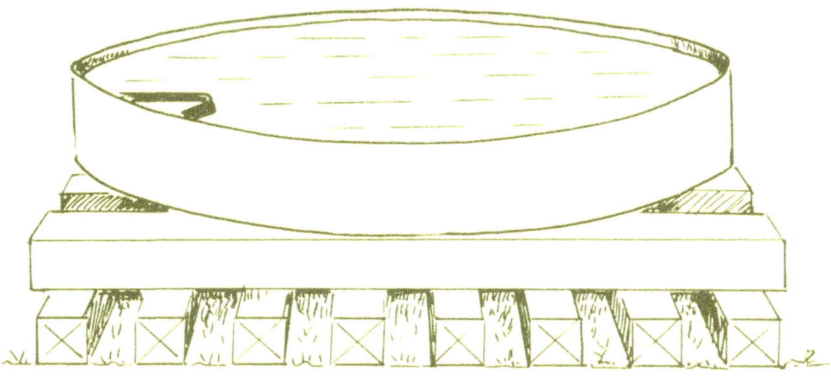

Figure 14. The Weather Bureau Class A Pan. The advantages of this design, recommended by the World Meteorological Organization, are standardization and ease of inspection for leaks. However, the exposure of the sides to radiation increases evaporation rates by about 40 percent. The readings therefore must be multiplied by an annual average "pan factor" of 0.7.

estimates of evaporation possible from routine meteorological readings. The basic requirements are thermometers in a ventilated instrument screen (maximum and minimum, wet bulb and dry bulb), a cup anemometer for daily wind-run at 2-meters height, and an estimate of solar energy. Solar energy may be measured with a sunshine-hour recorder, such as the Campbell-Stokes or Jordan instrument, or with a Gunn-Bellani distillation radiometer, which gives a fairly accurate estimate from a daily reading of a burette.(13) Where funds are more ample and technicians are skilled, an integrating thermopile radiometer may be used. With highly trained staff, Stanhill in Israel obtained good results from visual estimates of cloud cover.(16) Penman showed that a physically based estimate of open-water evaporation could be obtained from this standard meteorological data, which in some tropical countries has been recorded for many years.(8, 10) It is available for at least the past few years at all international airports.

In 1948, Thornthwaite published a method of estimating the "evapotranspiration" from a continuous canopy of green vegetation freely supplied with water.(19) He collected evaporation data from a wide area in North America and fitted arbitrary constants to relate evaporation to air temperature and day length. The method is easy to use and became popular with geographers, but it was at first applied uncritically to tropical climates for which the constants proved, not surprisingly, to be inappropriate. The results fell wide of observed tropical evaporation rates, and the method is now recognized as useful only within the conditions from which it was derived.

For routine calculations of water loss by evaporation, the Penman equation, with some useful refinements by others, is widely used throughout the world. Penman estimated evaporation first from the classical aerodynamic equation of Dalton—the product of wind speed and the difference in vapor pressure between the water surface and the windflow. He added this to the evaporation estimate derived from the heat-energy balance (incoming radiation, both shortwave and longwave, less reflection, less longwave reradiation, less heat storage in ground and vegetation) (see Fig. 6). Adjustments are made for the reflection of sunlight by different vegetation canopies and for their aerodynamic roughness. A weighting factor is used to adjust for seasonal differences in the relative effects of energy supply and ventilation. Where computer services are available, programs for the calculation of the Penman estimate of evaporation are available in all of the major systems and computing languages.

The automatic weather station is a further advance, which is now routinely used in some tropical watershed studies. The station is a mast bearing sensors that record electrically onto magnetic tapes of a data logger, which is a compact electronic device.(17, 18) The tapes are then transcribed by a compiler, and the calculations are done by a small computer. Enthusiasts for sophisticated equipment should remember that it is dependent on well-trained field staff and good maintenance. An automatic rain gauge, unless it is inspected daily, is even more susceptible than the simple rain gauge to a tropical species of wasp that seals its eggs in the rain gauge funnel with a generous plug of mud.

Changes in storage

Deep, freely draining soils are a characteristic of much of tropical Africa and tropical America and parts of Southeast Asia. The root depths of trees and grasses in such soils range from 3 meters to more than 5 meters. Within the root range, these soils have storage capacity for water, at tensions available to plants, that can exceed the total annual discharge of the stream.(12) Land uses involving changes of vegetation can therefore have strong and direct effects on the behavior of streams. Direct measurement of water-storage changes in the root range has been successfully developed as a routine in research on tropical watersheds.(13, 15) Deep stone-free soils are convenient for direct soil sampling, for deep installation of electrical tensiometers, or for drilling of access holes for the neutron-probe type of soil-moisture meter. Measurements of soil moisture are necessary to develop water-balance estimates for research purposes, but they require too much staff and supervision for routine monitoring for watershed management.

Beyond the root range, estimation of storage changes may be direct only in the small proportion of watersheds that are underlain by continuous groundwater surfaces, which are measurable from wells or boreholes. To estimate the specific yield—that is, the volume change represented by a rise in the well level—measurements are needed of the pore space or voids of the rock strata in which the water is stored.

Such direct measurement is rarely possible because the water table pattern is usually more complex in the streamsource areas, with groundwater lying in irregular aquifers of sand, gravel, or porous rock. As when a bath is emptied, the water drains most rapidly when the system is full, so that by plotting the recession curve as the streamflow dwindles, the depletion of the aquifer may be followed. Temporary increases due to rainfall events must be separated from the recession curve. The level at the end of the water year is then compared with that at the beginning to estimate the gain or loss in storage of groundwater (Fig. 15).

Changes in land use

Watershed management authorities in the tropics are liable to receive conflicting, but trenchant, advice from interested parties about the probable effects of proposed changes in land use. In the decades of rapid economic expansion that followed the Second World War, much of the technical advice was based on evidence from temperate-zone environments and economies, which had little relevance to tropical problems. Sufficient evidence has now been gathered in the tropics to permit technical diagnosis of the physical and biological conditions. It is therefore often possible to give useful estimates of the probable effects on streamflow and soil stability of major changes. These are from natural forest to arable cropping, grassland, or tree plantations.

For the first half of the 20th century, watersheds were considered to have such complex hydrological pathways that the only integrating factor was the streamflow. Rainfall and streamflow were therefore measured in pairs of similar and adjacent valleys for calibration periods of 5 to 10 or more years. An experimental change in land use (usually the clear felling of forest) was then made in one valley while the other served as a control. Observations were continued for a further 5 to 10 years.

This method was developed by U.S. forest hydrologists from their first study at Wagon Wheel Gap in Colorado at the beginning of the century, in which clear felling of a pine forest in-

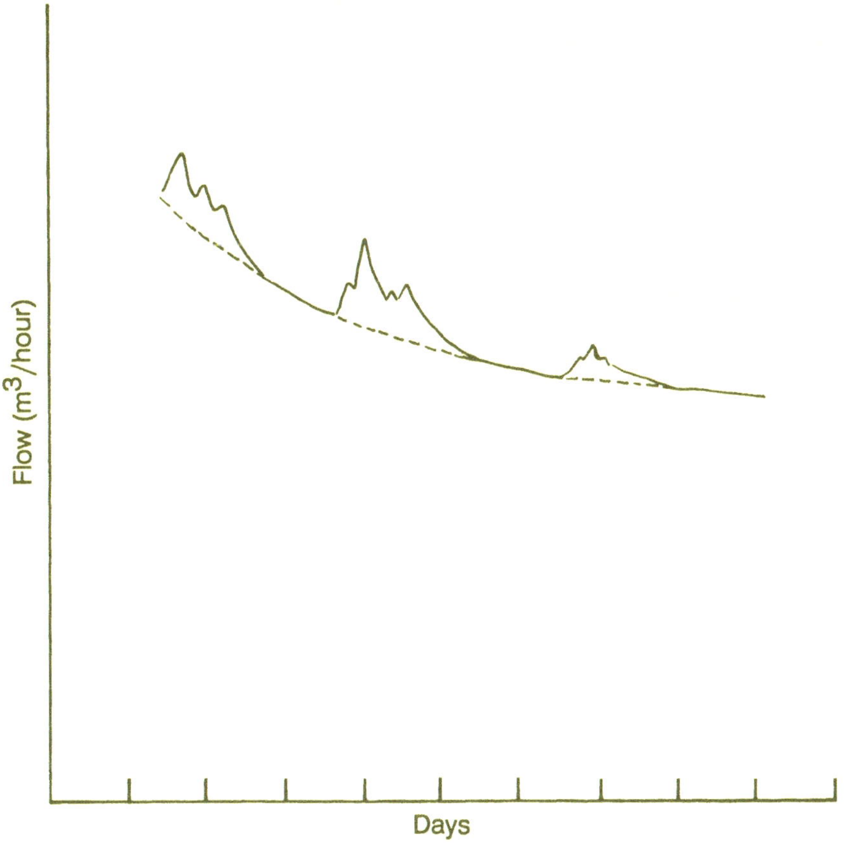

Figure 15. The groundwater recession curve. The depletion of groundwater storage may be traced by eliminating the effects of rainfall incidents during the dry season.

creased the water yield by 17 percent. In 1934, the U.S. Forest Service established the Coweeta Experimental Station to conduct studies on 40 small forested watersheds. Hibbert collected and compared the results of 39 such experiments in several countries.(4) The results showed qualitatively that the felling of forests increased water yields, whereas the planting of trees reduced them; but the results varied widely and were quantitatively unpredictable.

Many more such experiments were set up on the assumption that each would at least be valid for local prediction. The total had reached 94 when they were reviewed recently by Bosch and

Hewlett.(2) Of these, 67 were in the U.S.A., 10 in South Africa, 5 in Australia, 5 in Japan, 2 in Kenya, 2 in New Zealand, 2 in Madagascar, and 1 in Canada. They included rainfall regimes ranging from 400 millimeters to 2400 millimeters per year. The results showed such a wide scatter that they clearly have no general predictive value.

Budgeting for both water and energy

If measurements are confined to rainfall and streamflow, the difference between them is explained by evaporation, by storage changes, and possibly by deep seepage. A more intensive method of watershed study became possible through the development of the Penman equation, which allows direct estimation of water losses by evaporation. It was applied immediately to watersheds, with encouraging results. The estimate of evaporation accounted for the differences between rainfall and runoff for the 40 river basins in England for which long-term meteorological records were available.(9)

The first application to tropical watersheds began in 1955 in East Africa, where practical land-use problems were studied by full-scale conversions in comparison with undisturbed control valleys. In Kenya, tall rain forest was clear felled, and a fully equipped tea estate was established; in another area, a bamboo-forested valley was converted to pines. The water and energy balances established during the first 3 years have been verified during two subsequent decades of continuous operation.(1, 3, 13, 15)

Establishing a tea estate. Meticulous planning by soil conservation engineers and implementation by competent management have successfully replaced the hydrological controls of a dense tropical forest by a thriving plantation that includes a tea factory and housing and employs a large labor force. The main features are:

- Contour planting
- Hillside ditches (narrow-based terraces) leading runoff to prepared waterways
- Careful road construction with grassed banks
- Strips of original forest left to protect the steep stream banks

- Construction of a small storage pond on the stream to supply water to the factory

The effects of the land-use change were monitored. During the forest felling, the peak flows increased, but control was rapidly regained (Fig. 16). The peaks remained sharper and were increased when the tea was pruned, but they were too small to be of practical significance. Soil erosion losses were rapidly reduced to a negligible level.

Water yields were temporarily increased, by about 14 percent, for the first 3 years after clearing. The difference diminished as the tea developed, and when summed during the first 15 years, the extra yield was 9 percent. When fully mature, the tea estate matched the water yield of the forested control valley. Both yielded an average of 800 millimeters per year from an average rainfall of 2100 millimeters.

Establishing softwoods. The study of land-use change from dense bamboo forest to pine and cypress plantations was a simpler operation. Preliminary measurements on the Forest Department's observation plots showed that for well-grown pine and cypress the interception of rainfall by the canopies and the depletion of soil moisture to 3-meters depth were similar to that of the surrounding bamboo forest.(11, 14) Because the watershed was a critical source of water for Nairobi, the development was monitored by studies on small-streamsource valleys, comparing *Pinus patula* with undisturbed bamboo.

The monitoring showed that during the first 3 years, in which the bamboo was cleared and pines were planted among food crops (the *taungya* or *shamba* method), the water yield was 16 percent greater than that of the bamboo forest on the control catchment. When the pines were 16 years old and 25 meters high, the water yield was equal to that of the bamboo. The pines had then developed a deep leaf litter, and the control of runoff was fully effective, both streams showing the same pattern of peak flows. Both catchments then yielded an average of 1130 millimeters per year from an average rainfall of 2250 millimeters. The softwoods are harvested 20 to 25 years after planting.

Because these combined energy and water-balance studies included estimates of storage changes in soil moisture and

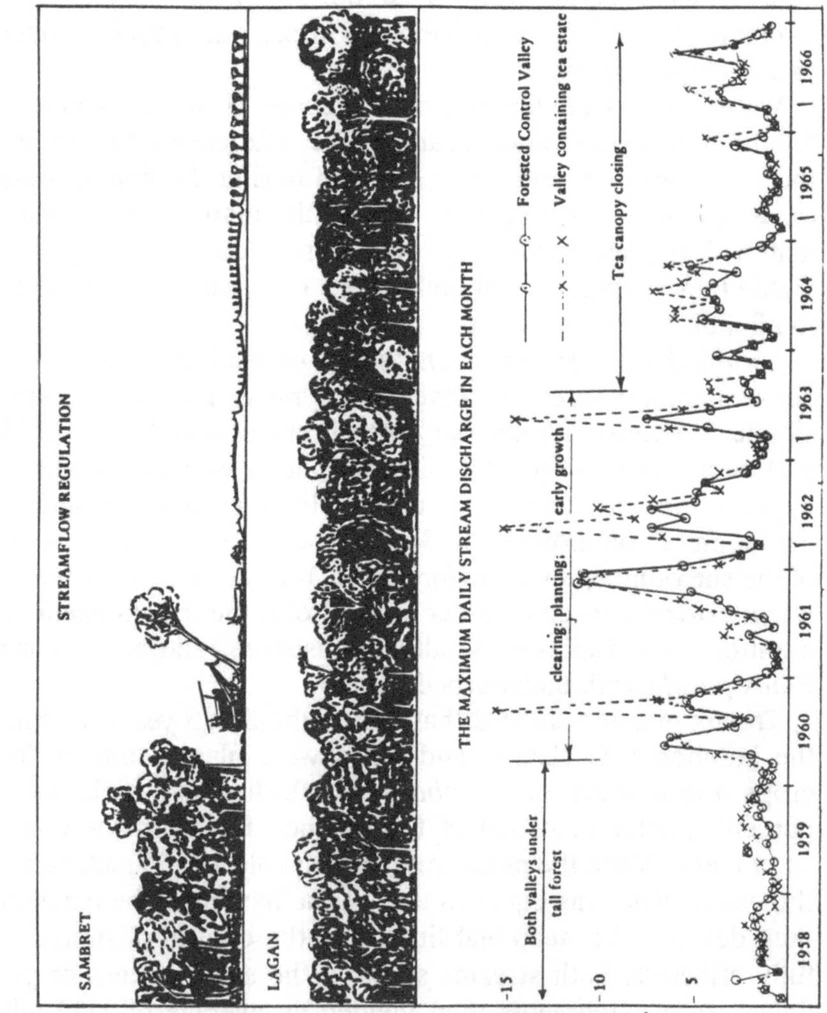
STREAMFLOW REGULATION
SAMBRET
LAGAN
THE MAXIMUM DAILY STREAM DISCHARGE IN EACH MONTH
Both valleys under tall forest
clearing: planting: early growth
Tea canopy closing
Forested Control Valley
Valley containing tea estate
mm depth over catchment area
15
10
5
1958
1959
1960
1961
1962
1963
1964
1965
1966

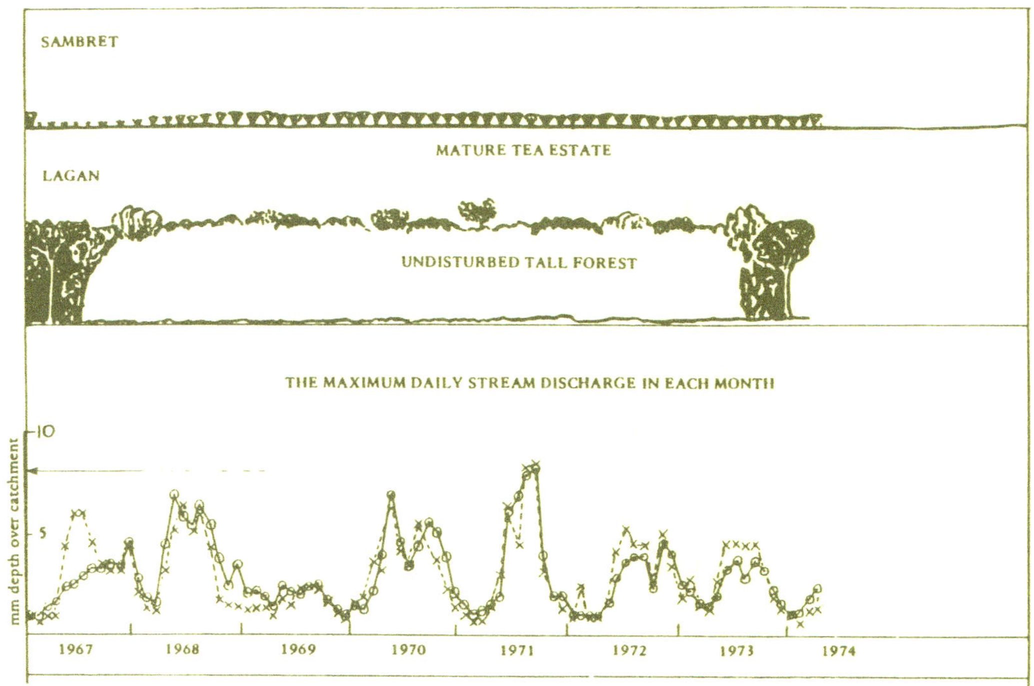

Figure 16. The hydrological effects of converting a tropical watershed from rain forest to tea plantings with careful soil conservation and watershed planning. The water-use and streamflow patterns of the mature tea estate (Sambret) were similar to those of the forested control catchment (Lagan).

Experimental conversion of a forested watershed in Kenya to a tea estate with careful conservation measures.

groundwater, they provided an independent estimate of all the terms in the hydrological equation. This provided a most valuable check as to whether the valleys were watertight, that is, whether all of the flow passed through the stream gauges. Out of the first 11 valleys equipped in the East African studies, two showed differences between rainfall and streamflow too large to be accounted for by evaporation or storage. In both cases, deep excavation revealed ancient buried streambeds along which water was bypassing the measuring weirs. These valleys were abandoned for study purposes. Lack of such checks is the most probable explanation of the wide scatter of results in the trials reviewed by Bosch and Hewlett.(2)

Models

Recent developments in computer technology have permitted rapid processing of the large volumes of data generated by watershed studies. Field experimentation on hydrological processes has been followed by their mathematical simulation or modeling.

Rainfall incidence and distribution, interception by vegetation, infiltration into soil, slow drainage movements through soils and rocks, transpiration and free evaporation from wet surfaces, storage changes, springflow, and finally streamflow can be simulated. All need checking against real-life field measurements from instrumented watersheds to secure validity for the models, which would otherwise remain hypothetical. Successful modeling of the East African watershed experiments has been achieved by Blackie, Edwards, and Clarke.(1) Such models permit useful exploration of the possible effects of proposed land-use changes. They can be based on existing data from routine measurements supplemented by relatively brief fieldwork. As with the aeronautical flight simulator, it is better to test for crash conditions on the computer than with a new aircraft.

Social and economic factors

There is now a scientific base for the development and management of tropical watersheds. Major investments by the governments of tropical developing countries in cooperation with the World Bank and other agencies are now designed after extensive field surveys by interdisciplinary teams of technical specialists. The foregoing examples of long-term, detailed studies of land use in East Africa confirm that well-watered tropical hill forests can be put to productive economic use without loss of hydrological control. The tragic scale on which such land is currently misused is due to lack of administrative organization and political guidance rather than to lack of well-established technologies for tropical land use.

The more difficult problems are presented by the inhabitants of the watershed and their livestock, for which quantitative data are often very difficult to compile. Where there is road access and adequate administration, counts of people and animals should be available, but in the more remote streamsource areas of hill watersheds, where human and animal impact may be the most critical, reputed numbers are often widely inaccurate. Aerial photographs are valuable tools, especially for assessing the size of cultivated areas, the extent and density of tree cover, and the incidence of soil erosion in grazing areas. Estimates of population may be based on counting the number of family holdings and following up with a ground-truth reconnaissance.

Socioeconomic surveys of the many variables, such as land-tenure systems, access to markets, transport for produce, availability of improved seeds and fertilizers, access to credit, and extent of indebtedness, can all derive some help from science in the form of statistical design and analysis of sampling. The most difficult variable to assess, which may be the ultimate determinant of the outcome of programs for improvement of land use, is the degree of political commitment to the objectives by those who have authority to act. Regrettably, science can offer no help in this problem.

References

1. Blackie, J.R.; Edwards, K.A.; and Clarke, R.T. 1979. Hydrological research in East Africa. *East African Agricultural and Forestry Journal* 43 (Special Issue). 313 p.
2. Bosch, J.M., and Hewlett, J.D. 1982. A review of catchment experiments to determine the effect of vegetation changes on water yield and evapotranspiration. *Journal of Hydrology* 55:2–23.
3. Edwards, K.A., and Blackie, J.R. 1981. Results of the East African catchment experiments (1958–74). In *Tropical agricultural hydrology,* ed. R. Lall and E.W. Russell. Chichester: John Wiley.
4. Hibbert, A.R. 1967. Forest treatment effects on water yield. In *Proceedings of the international symposium on forest hydrology,* 527–543. Oxford: Pergamon Press.
5. Huff, F.A. 1955. Comparison between standard and small orifice rain gauges. *Transactions of the American Geophysical Union* 36:689–694.
6. McCulloch, J.S.G. 1962. Measurements of rainfall and evaporation. *East African Agricultural and Forestry Journal* 27 (Special Issue):64–67.
7. Mills, H.R. 1900. The development of rainfall measurement in the past 40 years. *Simons British Rainfall* 40:23–45.
8. Penman, H.L. 1948. Natural evaporation from open water, bare soil and grass. *Proceedings of the Royal Society of London (A)* 193:120–145.
9. Penman, H.L. 1950. Evaporation over the British Isles. *Quarterly Journal of the Royal Meteorological Society.* 76:372–383.

10. Penman, H.L. 1963. *Vegetation and hydrology.* Technical Communication no. 53. Farnam Royal, U.K.: Commonwealth Agricultural Bureaux.
11. Pereira, H.C. 1952. Interception of rainfall by cypress plantations. *East African Agricultural Journal* 18:1–4.
12. Pereira, H.C. 1959. A physical basis for land use policy in tropical catchment areas. *Nature* 184:1768–1771.
13. Pereira, H.C. 1973. *Land use and water resources.* Cambridge: Cambridge University Press.
14. Pereira, H.C., and Hosegood, P.H. 1962. Comparative water use of softwood plantations and bamboo forest. *Journal of Soil Science* 13:299–312.
15. Pereira, H.C.; McCulloch, J.S.G.; Dagg, M.; Kerfoot, O.; Hosegood, P.H.; and Pratt, M.A.C. 1962. Hydrological effects of changes in land use in some E. African catchment areas. *East African Agricultural and Forestry Journal* 27 (Special Issue). 131 p.
16. Stanhill, G. 1965. A comparison of four methods of estimating solar radiation. In vol 25 *Proceedings of the Montpellier symposium: Arid zone research.* Paris: UNESCO.
17. Strangeways, I.C., and Smith, S.W. 1985. Development and use of automatic weather stations. *Weather* 40:277–285.
18. Strangeways, I.C.; Turner, M.; and Insell, W.S. 1980. A simple hydrological instrument system. *WMO Bulletin* 29(1):16–19.
19. Thornthwaite, C.W. 1948. An approach towards a rational classification of climate. *Geographical Review* 36:55.
20. U.S. Department of Agriculture. 1979. *Field manual for research in agricultural hydrology.* Agricultural Handbook no. 224. Washington, D.C.
21. U.S. Geological Survey. 1952. *Water loss investigations: Volume. 1. Lake Hefner studies.* Circular 229. Washington, D.C.

4
The role of forests in watershed management

The effects of forests on climate

The worldwide evidence that high hills and mountains usually have both more rainfall and more natural forest than do the adjacent lowlands has, historically, led to confusion of cause and effect. Although the physical explanations have been known for more than 50 years, the idea that forests cause or attract rainfall has persisted. The myth was created more than a century ago by foresters in defense of their trees. Then, as indeed today, the dwindling forest reserves were under attack by the fires, axes, and browsing livestock of the local people. In government councils, with their endless political maneuvering for new revenue sources, forests were no less at risk. The myth was written into the textbooks and became an article of faith for early generations of foresters.

The physical explanation is that winds drive warm, wet airstreams from the tropical oceans inland and up the slopes into cooler altitudes. There the water vapor condenses to rain or sometimes to hail. In the absence of hills, coastal deserts persist. Inland lakes show the same effects. On the equator in East Africa, the air is moistened as it crosses Lake Victoria, but only a narrow strip around the shore gains extra rain and the plains are dry. When the airstream reaches the Kerichȯ highlands, it is forced upward and heavy rain occurs on most days of the year. As a result, tall rain forests cover the hills wherever soil conditions and temperatures permit.

The reception of rainfall and the control of floods

Forests contribute to the stability of watersheds by protecting the soil surface from the direct impact of intensive tropical rainstorms. Rainfall is intercepted by the forest canopy, and the foliage temporarily retains a very large area of water films. Additional "detention" storage is provided in organic litter on the forest floor, which may be 10 to 30 centimeters deep, and in the surface soil, which has a very high pore space resulting from the actions of roots and of soil fauna. Water rapidly infiltrates vertically into the subsoil and drains laterally into the stream. The overall effect is that forests smooth out the concentrated inputs from intense rainfall and release a more regulated flow to the stream channels (Fig. 17).

The desiccation that so often follows the unwise clearing of tropical forests in climates that have a severe dry season is not due to lack of rainfall but to misuse of the land surface. When intense rainfall is shed immediately as surface runoff and streamflood flow, the storage spaces in porous rocks and soil are not refilled. Springs dry up and rivers dwindle. The sun may heat bare soil surfaces to 70°C or more, and winds blowing across them become hot and dry, so that crops and pastures may be desiccated. All these apparent symptoms of drought are contradicted by the rain gauge. The hills are still there and the rain still falls, but the forest, which had received the storms and regulated the streamflow, has vanished (Fig. 18).

Water evaporated from the oceans is estimated to provide about 90 percent of the precipitation over the continents of Asia and North America. Only two tropical forests, covering the watersheds of the Congo and the Amazon, are large enough to have some effect on the moistening of major airflows. Attempts have been made to show that the Amazon forest does increase local rainfall, but the data are too sparse to be conclusive.(16)

Although forests occur where water is available to support them and are not themselves the cause of the rainfall, they do have important local effects on the climate, due to the high rates of evaporation from the foliage. Evaporation is increased both by the aerodynamic roughness of the canopy and, after rainfall, by the very large wetted surface area of the foliage. The high rate of

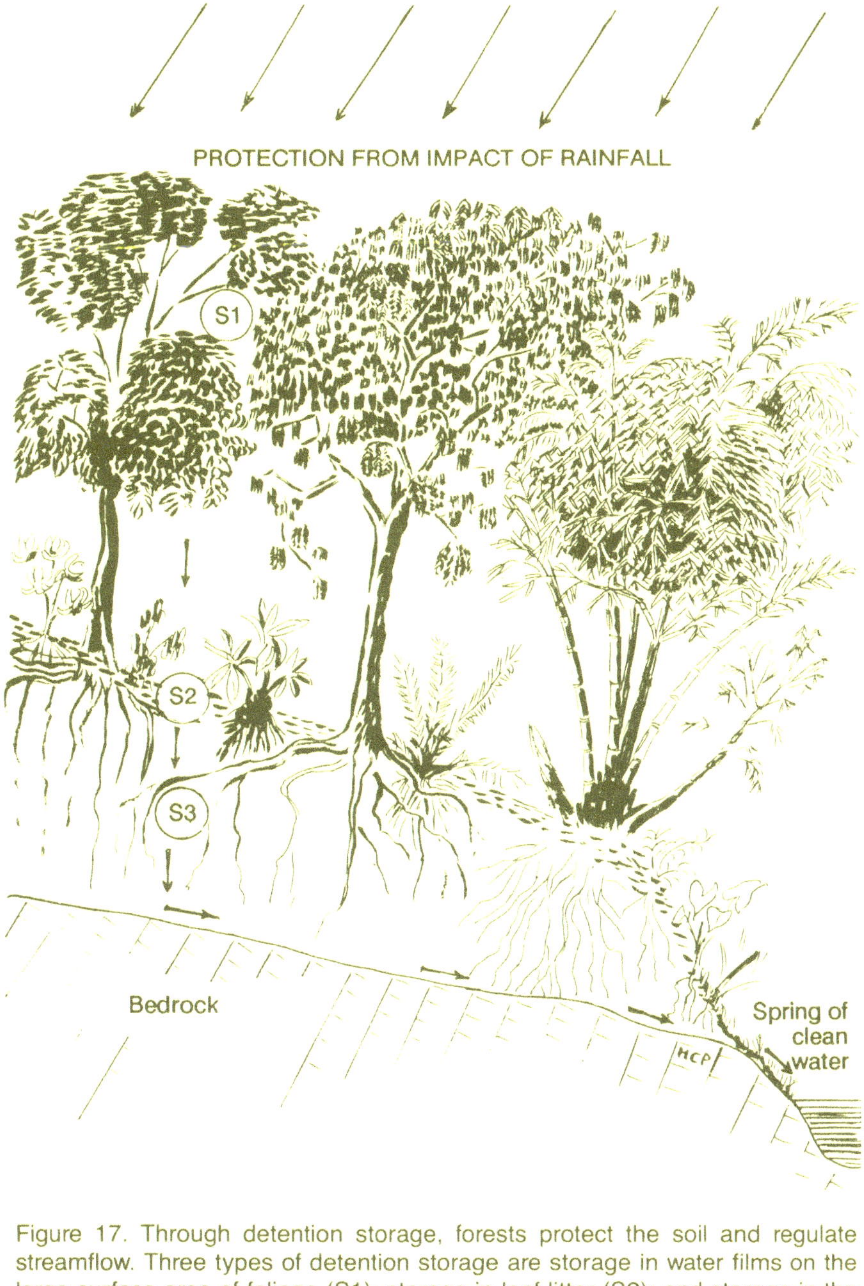

Figure 17. Through detention storage, forests protect the soil and regulate streamflow. Three types of detention storage are storage in water films on the large surface area of foliage (S1); storage in leaf litter (S2); and storage in the large soil pore space—root channels and burrows of soil fauna (S3).

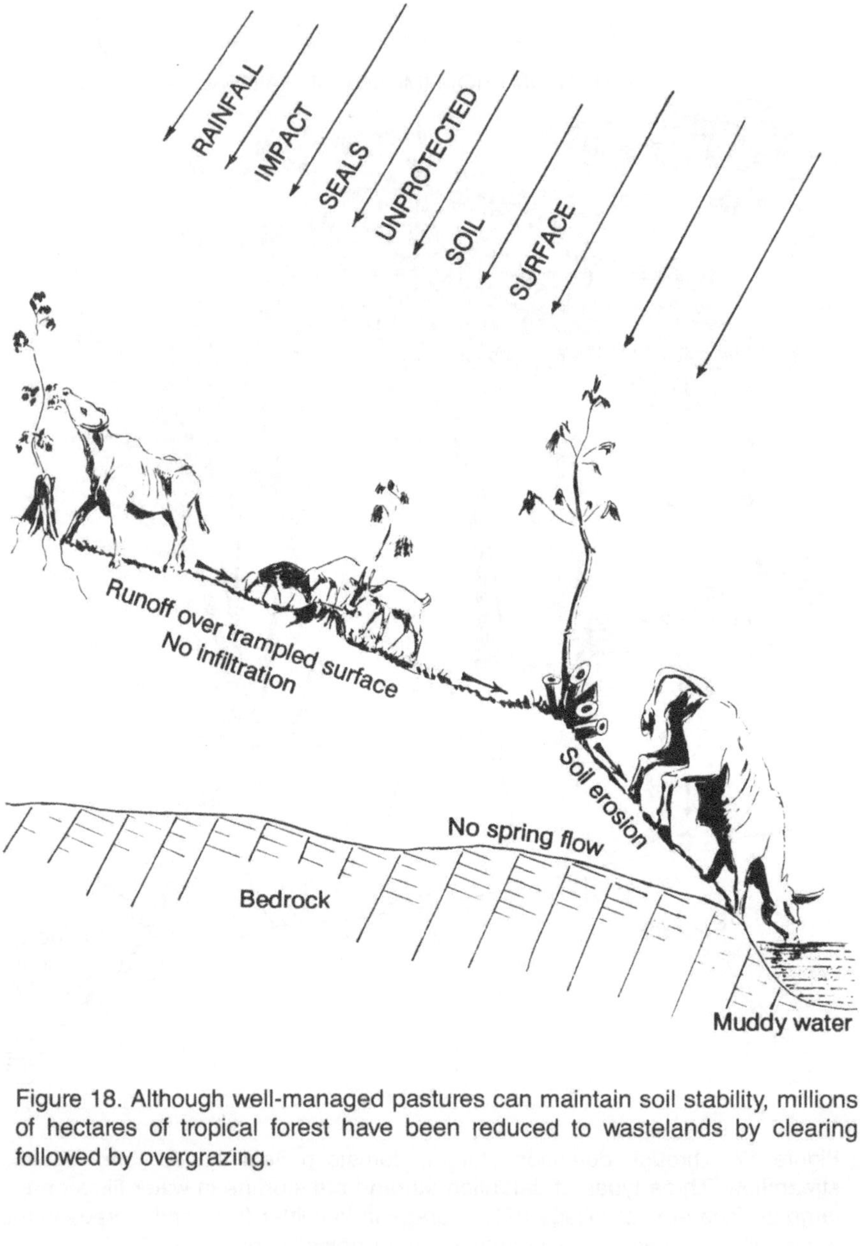

Figure 18. Although well-managed pastures can maintain soil stability, millions of hectares of tropical forest have been reduced to wastelands by clearing followed by overgrazing.

evaporation cools the forest and the air immediately above it. The result is the mist that is a characteristic feature of dense forests. In the U.K., the Institute of Hydrology employed tall steel towers that carried instruments to measure these processes both within and above a forest canopy. Scientists from the institute are now using similar techniques in the forests of the Amazon near Manaus. These studies indicate that 50 percent of the water vapor in the airstream above the canopy appears to have come from the forest(17, 19). The Amazon forests are so vast that their effect is comparable to that of an inland sea.

The energy balances of the East African watershed experiments (see Chapter 3) showed that in the dry season the heat flux to the atmosphere from tropical closed forest is about half of that from subsistence food crops and rough pasture.(13)

Given that both forests and heavy rainfall occur on steep slopes high in the watersheds, the ability of forests to delay torrent flow and to smooth the peaks of streamflow is of great practical importance. On flatter land, however, a dense growth of trees or bushes can impede streamflow, so that shallow drainage lines have to be cleared periodically, for example, of salt cedar in the U.S.A., of willows in New Zealand, and of a rich variety of lush vegetation in the tropics. In the temperate zones, the valuable role of forests in the regulation of floodflow has long been established. The evidence is well reviewed in the *Proceedings of the international symposium on forest hydrology*(18) and, for the U.S.A., in a U.S. Forest Service report by Anderson, Hoover, and Reinhart.(1)

Early quantitative evidence on the effects of forests on streamflow came from the southern United States, where a severely eroded watershed of 90,000 square kilometers was restored, beginning in the 1930s. The Tennessee Valley was an agricultural area impoverished by misuse of land. As an essential stage in the development of a series of storage reservoirs for hydroelectric power, productive land management was established.(20) For example, in Pine Tree Branch, a badly eroded subwatershed was stream gauged for 5 years before reforestation began. The surface drainage of the catchment area was reorganized to follow cutoff drains along controlled gradients. The development of gullies was checked by installing temporary brushwood structures and seeding fast-growing shrubs. Then the whole subwatershed was planted to pines. The trees grew well, and the effects on the stabilization of the watershed were most convincing. Comparison with initial measurements showed that the peak rates of torrent flow were reduced by 90 percent and the sediment load of the stream was reduced by 96 percent. Floods and soil erosion were eliminated, but the water yield of the valley was halved. The loss of water yield is the price paid for the survival of the reservoirs and hydropower investments.

Physical limits to flood control by forests

Flood regulation by forests has physical limits that are from time to time overwhelmed by storms of sufficient size and duration. One limitation is the depth of soil and its capacity for water storage. This fact was demonstrated in the early days of the Coweeta Hydrological Laboratory in the Appalachian Mountains by two pioneer American forest hydrologists, Charles Hursh and Marvin Hoover. A rainstorm of 200 millimeters at low intensity fell on two adjacent watersheds of similar topography that were already wet. One had an average soil depth of 2 meters. It produced a stormflow peak of 2.20 $m^3/sec/km^2$. The other had shallower soils averaging only 0.6 meter deep, and its stormflow peak was more than five times as large, 12.25 $m^3/sec/km^2$.(5, 6)

Another limitation is that even very large areas of forest are unable to control floodflows when they become saturated, as occurs during the monsoon in the eastern Himalayas and Burma, where up to 5000 millimeters of rain may fall in 6 months. Measurements made in the headwaters of the great Mekong River provide an example. At Chiang Saen in northern Thailand, the Mekong has been gauged for 20 years as it flows from an upper watershed of some 200,000 square kilometers, much of which is forested. The average flow is 2800 m^3/sec with 6 months of low flow and 6 months of high flow rising to a peak in September. The average of this annual peak is 46 times the average low flow. In 1966, the seasonal rise reached 12.3 meters.(10)

Although the forests may be overwhelmed by the monsoon, their role in the protection of the watershed becomes even more critical. By providing a physical obstruction that slows overland flow, by protecting the topsoil, and by strengthening riverbanks, forests minimize the transport of soil and thus moderate the main mechanism of watershed degradation.

The transport of soil

Measurement of the quantities of soil transported in large floods is both difficult and costly. Such data are scarce in tropical countries but are available from major land areas under subsistence agriculture in China. The transport of soil by floods is of major national concern in China, where vast quantities of silt are deposited in the slow flowing lower reaches of rivers.(12) The sediments block the river channels and spill floods across the plains. The populations, which concentrate in the ricelands of the valley bottoms, strip trees and grasses and topsoil from the hillsides and even uproot young pine trees. This misuse of land has caused spectacular erosion. A long-term study by the Academia Sinica has estimated 20 million hectares in the watershed of the Huang Ho (Yellow River) to be damaged.(12) As a direct result, irrigation engineers struggle with floods bearing more than 50 percent by weight of solid material in suspension. Similarly, the forested upper reaches of the Yangtze River have been denuded since the 1950s, leaving only 3 percent of forest cover. Flood damage has become so severe that the government has begun a reforestation program with a target of 70 million hectares by the year 2000. Although such reforestation cannot prevent the floods caused by major atmospheric events, it will reduce the destructive effects both by damping the peak flows and by reducing erosion and the transport of soil and debris.

Logging of natural forest

Forests are a major source of foreign exchange earnings for at least 15 developing countries. Their timber exports total U.S.$8 billion a year. Although forest cover should be retained on steep slopes, particularly when high in the watersheds of important rivers, this does not mean that the forests should not yield timber or fuelwood. The essentials for watershed management are that the harvesting must be planned and controlled to minimize damage and that the forest must be protected from invasion along the tracks and over the bridges built by the timber extractors. As cities grow, the illicit firewood contractor with large trucks and rapidly assembled labor gangs is a major threat to the survival of a forest opened up by selective logging.

Because only a few species are sufficiently well known to the export trade, large-scale waste occurs when tracks are cut through dense forest to reach the few chosen trees. The dipterocarp forests of Southeast Asia are severely damaged by logging because only one-fifth of the standing timber is harvested, while more is pushed over and left as tangled wreckage of vegetation interlaced by a maze of wheel tracks and skid furrows. On steeply sloping land under heavy rainfall, the tracks initiate serious soil erosion. In West Africa, where only 10 out of 150 species are considered salable, the same wasteful processes have been traditional.

Opportunities for major improvements depend on changing the selected-species management philosophy derived from Europe. This reorientation may occur as timber consumption within developing countries increases. Because a far wider range of species is locally acceptable, simultaneous harvesting of mixed tree stands is economically feasible. By greatly shortening the access tracks, damage to soil and regrowth is minimized.(15)

The damaged land often gives poor regrowth. An FAO/UNEP study of the forests in 76 developing countries showed that out of 169 million hectares of logged-over land, only 42 million hectares are being managed and protected. Most of the managed area is in India, where the federal and state forest departments have some traditional as well as legal authority after a century of experience.(3)

Even when managed, the productivity of logged-over land is low, producing about 1 m^3/ha/yr of timber as compared with 5 to 20 m^3/ha/yr when trees are replanted. With regrowth from natural forest, protection must be maintained for 40 years following harvest in order to develop a second crop of the harvested species. But when logging opens the canopy, vines and fast-growing pioneer species may overwhelm the desirable seedlings so that a second crop is uncertain. Enrichment planting of marketable species has proved to be too expensive. Forestry studies in many tropical countries spanning a half century have failed to develop economic methods of sustained timber yield of slow-growing hardwood species in natural forest. New developments in tissue-culture techniques for the rapid reproduction of elite specimens of tropical hardwoods such as the West African *Triplochyton scleroxylon* appear encouraging.(8)

Protection of steep upland watersheds must therefore depend on government policy and resources for land management. Forest departments are usually compelled to concentrate their limited forces on the protection of the remaining salable stands of timber in virgin forests. In most countries, governments have given higher priority to funding the planting of trees for industrial forest products such as timber, pulpwood, and fuel. This policy is indeed essential to reduce the rate of depletion of natural forests, but an active policy is also needed to protect forests on steep slopes.

The aftereffects of commercial logging vary with topography. On mild slopes, the damage done by commercial logging is short-lived if the cutover forest is protected because loggers remove only about 10 percent of the trees. The rest of the trees lost are removed for fuel or destroyed by forest fires. Rapid vegetative growth restores soil protection. On steep slopes, however, when the logging is completed, drainage must be diverted from the tracks to prevent the development of gullies. By minimizing extraction costs at the expense of the land, contractors in the U.S.A. and in many tropical countries have done severe and widespread damage.(9, 14) Examples of such practices are bulldozing of excessively steep tracks for dry weather use, filling drainage lines with soil to avoid installing culverts, dead-snigging or dragging logs by tractor without raising the butts, and winching downhill to loading points, thus creating radial patterns of furrows that concentrate runoff and initiate gullies (Fig. 19). All these sources of damage may be avoided by simple techniques that are well established in more developed countries and have been applied successfully in tropical areas such as Queensland in Australia.(4)

Logging contracts should therefore exclude steep slopes and

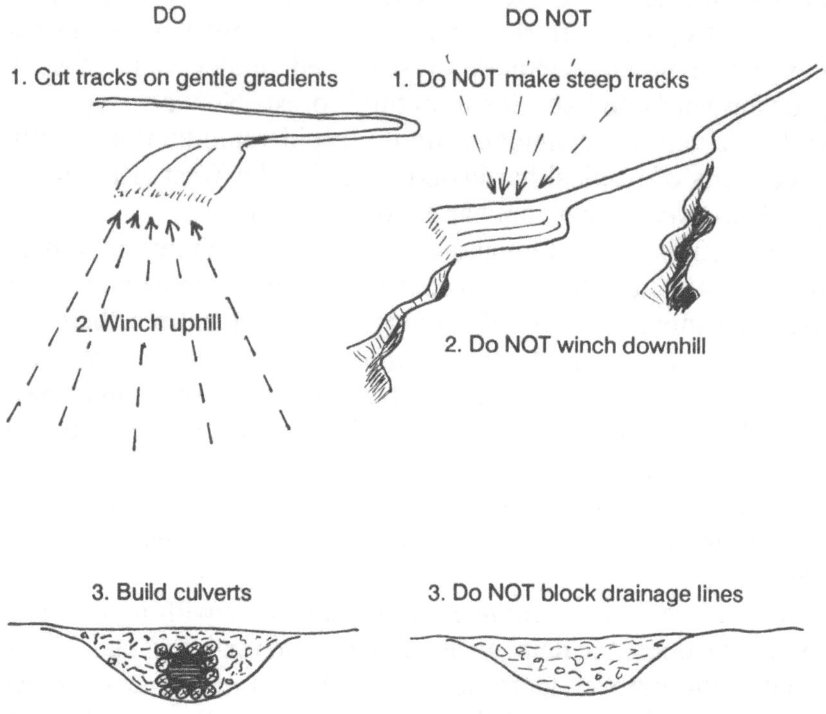

Figure 19. Minimizing damage to soil and water resources from logging on slopes.

should specify other precautions such as the maximum gradients of haulage roads, winching uphill rather than downhill to log landings, and the building of culverts (Fig. 19). Forest departments need sufficient staff and vehicles to monitor contracts. Perhaps a greater need is political commitment in support of such control, because logging contractors tend to be wealthy and influential citizens.

Yet even a carefully designed road system will increase the rate of runoff from a watershed. In a summary of some five decades of research at the Coweeta Hydrological Laboratory in steep terrain with 2000 millimeters of rainfall, Douglass reported that logging under good management increased stormflow by 17 percent.(2) The size of the stormflows declined rapidly with regrowth.

For small upland communities in the tropics, where employment opportunities are scarce, labor-intensive methods of harvesting are appropriate. Processing at stump by pit-sawing is a traditional practice that has been unpopular with foresters because it is used by poachers. Where hill communities are formally involved in the management of local forests with the guidance of trained foresters, as in the evolving system of *panchayat* (community) forests in Nepal, these simple technologies are desirable. The sawn timber is carried by hand down to the villages so that damage by logging vehicles is eliminated. Removal of scattered over-mature trees in this way need have no adverse effects on soil stability or water control (Fig. 20).

Dry tropical forests

Watershed management is more difficult in tropical climates that have severe dry seasons because fire becomes a major hazard. In the dipterocarp forests of Southeast Asia, fire control is a dominant problem because these deciduous forests shed large volumes of leaves and twigs. In Thailand people deliberately set fire to the hill forests every year. The smoke that covers the hills prevents the use of the watchtowers. Although the main species of trees are fire resistant, so that grown trees survive, the seedlings are destroyed and the soil is left bare. Rainfall impact causes sheet erosion, which increases the sediment carried by the rivers. Educating the public about the connections among heedless fire spreading, sedimentation of riverbeds and canals, and greater flood damage needs to be a major factor of watershed management.

Much woodland savanna survives in the tropical highlands of Africa because human settlement has been restricted by the lack of both surface water and potable groundwater during the long dry seasons. With annual rainfall totaling 1000 millimeters and evaporation totaling 1250 millimeters, there is no surplus to recharge aquifers and to maintain springflow. Open woodlands of acacias in Kenya's bushland; of *Brachystegia* species in the *miombo* of Tanzania, Zambia, and Malawi; and of *Colophospermum* species in the *mopane* of Zimbabwe, Botswana, and Mozambique all play an extremely important role in stabilizing of hillsides by their networks of roots. However, their open growth

Figure 20. Labor-intensive methods are suitable to community forestry in remote watersheds. Mature trees can be harvested for local use, with minimum damage to soil and water stability, by pit-sawing and carrying out the lumber.

permits grass and shrubs to proliferate, presenting a serious fire hazard. Even in areas that have a small human population, herders and hunters ensure annual burning either deliberately to renew grass growth or to drive out game or accidentally when smoking out bees' nests. Annual fires in combination with overgrazing by livestock cause widespread sheet erosion and intensified torrent-flow patterns.

Dry woodlands have proved to be very fragile. In the Sahel, they have rapidly degenerated into wastelands. Good watershed management in dry woodlands requires control of livestock numbers, the rotation of grazing, and some prescribed early burning so that the grass is fired early in the dry season while the soil is still moist, which minimizes damage both to grass roots and to trees. (The uses of trees by agricultural settlements in drylands are discussed in Chapter 6.)

The use of trees in dryland watersheds

The water use of trees is a sharply debated issue of watershed management in regions of the tropics where dry seasons are severe and annual evaporation exceeds rainfall. Although trees are valued for fuel, fodder, and maintenance of watershed stability, the water they consume affects the availability of supplies for people and animals.

The main facts are now well established. Infiltration of rainfall at the soil surface is essential to recharge groundwater. This process can often be enhanced by reforestation of steep slopes that have been stripped of their original cover. Where soil erosion has already started, the digging of shallow furrows on the contour is widely used and successful in areas where manual labor is freely available.

On steep bare loess hillsides of Gansu province in northern China, large-scale planting for fuelwood is done on hand-built terraces only 1 meter wide. An early yield is obtained by sowing *Astragalus adsurgens,* a fast-growing shrub. In Shaanxi province, because of severe erosion, the soil surface is being experimentally stabilized by aerial seeding before planting of trees. Of 32 species seeded from the air, the best performing was alfalfa (*Medicago sativa*), sown at 2.2 kg/ha. Because alfalfa can be grown successfully in tropical highlands, this method is worth wider study.

Trees, however, should not be planted across marshy ground in springsource areas. Tree stands have a lower reflection coefficient than a grass or reed cover. They therefore absorb more radiation, which is used for evaporation while water supplies are available. Also, the foliage of trees is exposed to more ventilation, and by increasing aerodynamic roughness, the foliage itself further increases evaporation. A practical method used in the past to combat malaria in East Africa was to plant *Eucalyptus robusta*

A watershed in China's loess hills (in the province of Gansu) has been rehabilitated by large-scale planting of the woody shrub *Astragalus adsurgens.*

In the loess hills of Shaanxi province, China, formerly denuded slopes (above) and gullies (below) are covered with alfalfa 3 years after aerial seeding. The alfalfa stabilizes the soil prior to planting trees. *Source:* Northwest Institute of Soil and Water Conservation, China.

directly into marshlands in order to lower the water table and to dry the surface.

Trees should not be planted so close to stream banks that their roots can reach the water table. Where this has been done, measurements of water levels taken both in mornings and evenings will show a diurnal variation, with the levels falling during the day and recovering during the night (Fig. 21). Where water yield is critical, some saving may be obtained by cutting trees and shrubs along the river margin. These plants will be natural water-loving species (phreatophytes) whose roots are well adapted to wet soils. The grasses and forbs that replace them will transpire perhaps 20 percent less.

This recommendation may appear to conflict with the policy of leaving the natural vegetation to protect the soil on steep steam banks. The difference is illustrated by Figure 21. In steep topography, the streams have usually cut down to rock and there are no high water tables to be tapped by the stream-bank vegetation. No rule of thumb can be formulated for the width of the margin to be left unplanted, or even kept cleared, to increase water yield. The distance must depend upon the extent of the water table. In South Africa, a legal minimum 30-meter width of stream bank has to be left unplanted.

Water yield becomes a dominant concern for watershed management in tropical climates that receive less than 500 millimeters of rainfall a year. Characteristically violent convectional rainstorms cause spate flows in streams and severe sheet and gully erosion in misused lands. Trees that have evolved in such harsh environments have widespread surface root systems. Researchers in Zimbabwe have found *mopane* trees 15 meters high that have roots covering a circle 50 meters in diameter. Figure 22 shows a similar root habit of *Acacia senegal* in the Sudan. Such trees stabilize topsoil and provide both shade and wind protection as well as yielding fuelwood, forage, and, sometimes, resins. However, where a watershed is, for instance, supplying a city, the water use of trees may well be of critical importance. In Australia, for example, Melbourne's population of 2.5 million is supplied from 121,000 hectares of forested mountain watersheds, under annual rainfall of 1600 millimeters. A recent study of 4 years of records from nine watersheds showed a significant rela-

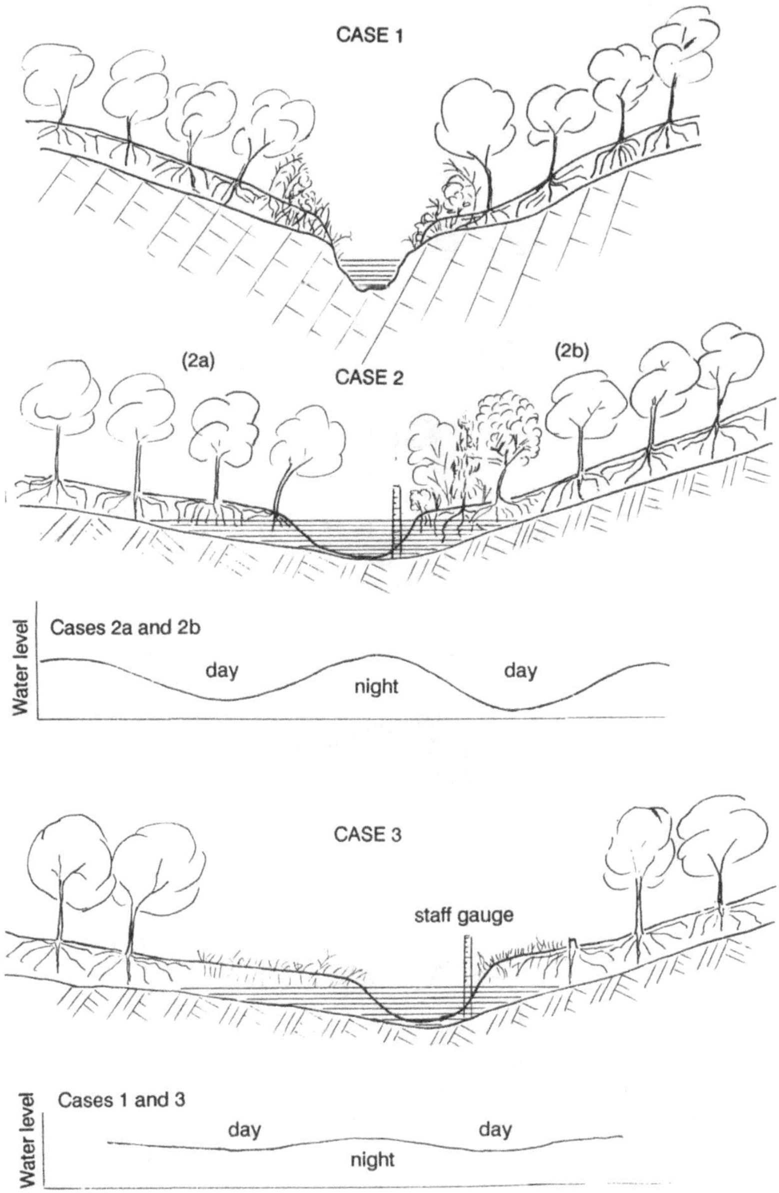

Figure 21. Stream-bank management for water conservation: (1) On steep slopes, leave a protective strip of natural vegetation; the stream channel is incised into rock so roots cannot reach the water. (2) On mild slopes with a high water table near streams, do not plant trees near the stream bank (2a) and do not permit water-loving trees and shrubs to grow near the stream bank (2b). (3) On mild slopes, short vegetation on the stream bank has little effect on water use.

Figure 22. *Acacia senegal* grows widely spaced. The root systems exploit large areas of surface soil and also seek water at depths of 3 to 5 meters. *Source:* From a study by G.E. Wickens (1984).

tionship between water loss (using rainfall-runoff) and the basal areas of their tree populations. Watersheds with mature forests of very large mountain ash (*Eucalyptus regnans*), with 20 trees per hectare, used from 100 to 200 mm/yr less water than watersheds with regrowth forests having 400 trees per hectare.(7)

In near-desert conditions of 300 millimeters of rainfall or less, highly adapted trees have an important stabilizing effect against wind erosion. Many areas in African ecologies dominated by the drought-resistant acacias such as *Acacia seyal, A. senegal,* or *A. tortilis* have suffered severe wind erosion when trees have been felled for firewood. Their reestablishment is an important means of soil stabilization. In Asia, *Prosopis cineraria* and *P. juliflora* are used for the same purpose. Under such severe climates, underground water resources are recharged by local concentration and infiltration of surface runoff rather than by a general absorption of rainfall. Large, widely spaced trees therefore do not reduce water supplies unless they are growing in drainage lines where their roots can reach the small areas of groundwater.

Trees at the seacoast

Forests have an important role right to the seaward end of the watershed. In Bangladesh, for instance, in the heavily populated delta of the Ganges and Brahmaputra, the flat coastline has in the past been protected by dense forests of mangroves growing in shallow water. These forests absorbed part of the energy of the waves created by the severe typhoons that afflict this coastline. In the 1960s, the mangrove forests were heavily cut and largely destroyed. When a major typhoon struck in 1970, some 150,000 Bangladeshis were drowned. The mangroves are now being reestablished and protected. If well managed, they can both provide protection and be a source of building poles, fuelwood, and charcoal of high calorific value.(11)

Sand-dune stabilization

In Senegal, efforts to stabilize sand dunes provide a good example of international cooperation and of long-term government commitment to the prevention of desertification. Along the coast for about 200 kilometers north of Dakar, there is a very productive and heavily populated zone of lowland agriculture. A natural cover of acacia trees and grasses maintained the stability of the coastal dunes until population growth led to intensified fuelwood cutting and overgrazing. As the cover was lost, wind-driven sand began to encroach on the farmlands.

Government efforts to control the dunes began in 1925, but by 1970 only some 20 kilometers had been reforested. From 1973 to 1982, a combined effort by UNDP/FAO, the World Food Program, and the Canadian International Development Agency stabilized a further 130 kilometers of dunes. In addition windbreaks of eucalyptus and cashew were planted, to a total of 637 kilometers. *Acacia albida* and *A. tortilis* were planted around homesteads to provide fuel and fodder. The costs of stabilization ranged from U.S. $50/ha for palisades woven from the twigs of a local shrub to $240/ha, of which reforestation cost $175/ha.

In spite of this technical success, the long-term issues remain in the balance. If population growth continues at its present rate, the whole destructive cycle could be repeated unless the farming communities can be persuaded to control their livestock and to protect trees planted in the dunes from felling for fuel.(21)

References

1. Anderson, H.W.; Hoover, M.D.; and Reinhart, K.G. 1976. *Forests and water: Effects of forest management on floods, sedimentation and water supply.* PSW18. Berkeley, California: Pacific Southwest Forest and Range Experiment Station, U.S. Forest Service.
2. Douglass, J.E. 1983. A summary of some results from the Coweeta Hydrological Laboratory. In *Tropical Forested Watersheds,* ed. L.S. Hamilton and P.N. King. Boulder, Colo.: Westview Press.
3. Food and Agriculture Organization and United Nations Environment Program. 1982. *Tropical forest resources.* Forestry Paper no. 30. Rome.
4. Gilmour, D.A. 1977. Logging and the environment, with particular reference to soil and stream protection in tropical rain forest situations. In *Guidelines for watershed management,* 223–235. FAO Conservation Guide no. 1, Rome: FAO.
5. Hursh, C.R. 1943. Water storage limitations in forest soil profiles. *Soil Science Society of America Proceedings* 8:412–414.

6. Hursh, C.R., and Hoover, M.D. 1941. Soil profile characteristics pertinent to hydrological studies in the southern Appalachians. *Soil Science Society of America Proceedings* 6:414–422.
7. Langford, K.J.; Moran, R.J.; and O'Shaughnessy, P.J. 1980. The North Maroonda experiment pretreatment phase: Comparison of catchment water balances. *Journal of Hydrology* 46:123–145.
8. Leakey, R.R.B.; Last, F.T.; and Longman, K.A. 1982. Domestication of tropical trees: An approach securing future productivity and diversity in managed ecosystems. *Commonwealth Forestry Review* 61(1):33–41.
9. Lundgren, B. 1978. *Soil conditions and nutrient cycling under natural and plantation forests in Tanzanian highlands.*Reports in Forest Ecology and Forest Soils no. 31. Uppsala: Swedish University of Agricultural Sciences.
10. National Energy Administration. 1977. *River gauging report.*Vol. 1. Bangkok.
11. Office of Technology Assessment. 1984. *Technologies to sustain tropical forest resources.*Washington D.C.: Congress of the United States.
12. Pereira, H.C. 1981. Rehabilitating eroded hill lands in the People's Republic of China. *World Crops*33:96.
13. Pereira, H.C., and McCulloch, J.S.G. 1960. The energy balance of tropical land surfaces. *Tropical meteorology in Africa: Proceedings of the Munitalp conference*. Nairobi: Meteorological Department.
14. Rice, R.M.; Rothatcher, J.S.; and Megahan, W.F. 1972. Erosional consequences of timber harvesting—An appraisal. *Proceedings of the symposium on watersheds in transition*. Urbana, Ill.: American Water Resources Association.
15. Roche, L. 1978. Community forestry and the conservation of plants and animals. *Proceedings of the eighth world forestry congress.* Jakarta: Eighth World Forestry Congress.
16. Salati, E.; Dall'Olio, A.; Matsui, E.; and Gat, J.R. 1980. Recycling of water in the Amazon Basin: An isotopic study. *Water Resources Research* 15:1250–1259.
17. Shuttleworth, W.J. et al. 1984. Eddy correlation measurements of energy partition for Amazonian forests. *Quarterly Journal of the Royal Meteorological Society* 110:1143–1162.
18. Sopper, W.E., and Lull, H.W., eds. 1967. *Proceedings of the international symposium on forest hydrology.* Oxford: Pergamon Press.
19. Stewart, J.B., and Thom, A.S. 1973. Energy budgets in pine forests. *Quarterly Journal of the Royal Meteorological Society* 99:154–170.

20. Tennessee Valley Authority. 1962. *Reforestation and erosion control influences upon the hydrology of the Pine Tree Branch watershed (1941–1960)*. Knoxville, Tenn.
21. World Resources Institute. 1985. *Tropical forests: A call for action*. Washington D.C.

5
Meeting the urgent fuelwood shortage in tropical developing countries

Watershed problems from fuelwood harvesting

Serious problems of watershed management are caused by the rapidly developing tropical fuelwood crisis. In developing countries, four-fifths of the wood harvested is used for fuel (Fig. 23). Throughout Africa, 76 percent of the energy consumed comes from fuelwood, as compared with 42 percent in Asia and 30 percent in Latin America. As populations grow and forests shrink, the traditional practice of gathering free fuelwood is causing worldwide difficulties.

China has the largest share of the problem with 350 million rural people suffering acute shortages of domestic fuel for 6 months of every year after the crop residues have been used up. In Jianxi province of southern China, I have seen villages where firewood collection from the natural forests involved walking for 2 days each way. Although China has large reserves of coal and gas, the distances in rural areas are far too great for economical distribution. Since 1949 the government has organized the planting of very large areas, although with rather variable success, so that China has more than 30 million hectares of planted trees. This area exceeds the total for 76 other developing countries listed by FAO/UNDP in 1982(2). China's present rate of planting is greater than the combined efforts of the rest of the world, yet

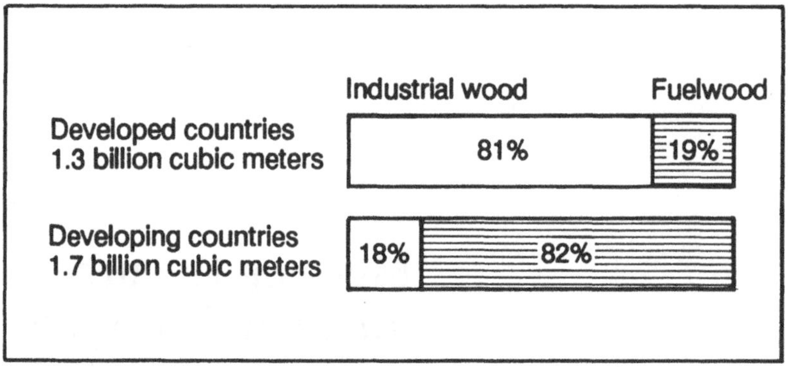

Figure 23. The use of wood as fuel in developing and developed countries. *Source:* FAO, *Yearbook of Forest Products 1983* (Rome: 1984).

the fuelwood shortage becomes still more acute as the remaining natural forests are destroyed.

Excluding China, South and Southeast Asia contain 30 percent of the world's population but only 2 percent of the world's known reserves of fossil fuels.(6) Major hydropower and nuclear power development requires such heavy investment that progress is very slow and is directed to urban consumers. Simple technologies suitable for rural areas have been evolved for small water-powered minihydel generators (Chapter 11) and for solar heating, but costs are major constraints to their wide adoption.

The problem is intensified by the drift of rural population to the outskirts of tropical cities, where charcoal is the common household fuel. Charcoal contractors drive ever farther afield as forest sources dwindle. Because it takes about 15 tons of fuelwood to produce 1 ton of charcoal in the traditional earthen kilns, the growing populations of cities add significantly to the rate of forest destruction.

This low efficiency of wood use is unnecessary. Modern metal kilns can improve the ratio to 5 tons of fuelwood per ton of charcoal, thereby reducing the demand on forests. Several good designs have been tested, demonstrated, and published by technical assistance organizations, and some, such as the Tropical Products Institute of the U.K., have deployed staff to assist with their adoption. Many universities and technical institutes offer more developments of the designs, but the far harder task is to teach their use in the field. In Uganda, charcoal production is combined with logging so that good use is made of the large volumes of treetops, crooked stems, and branches that are usually wasted. There metal kilns of the Casamance design have raised the annual production of charcoal from 200 to 63,700 tons in a decade.

Part of the fuelwood problem is the wasteful household practice of cooking over open fires. With technical assistance from Western laboratories, simple flue-type cooking stove designs have been provided for small pottery stoves that have been made locally by rural artisans. These stoves have greatly improved efficiency; tests in Nepal by the World Bank/FAO Community Forestry Project have shown savings averaging 562 kilograms of fuelwood per family per year. Social acceptance has been good, but major efforts will be needed to secure the general adoption of such changes in traditional village customs. More efficient stoves are particularly beneficial to women, who are primarily responsible for collecting fuel and transporting it over long distances. A study of fuel use by the 100,000 households in the Kathmandu Valley showed that the new stoves could save 92,000 tons of fuelwood a year, an amount equivalent to the yield from 14,000 hectares of fuelwood plantings.

In tropical America, more than 7.5 million hectares of closed forest were cleared between 1981 and 1985. Much of the lower slopes of the Andes are already denuded of trees. Fuelwood cannot be economically transported from the forests of the Amazon: local shortages must be met by tree planting. Peru and Brazil have successful large-scale forest plantations.

An FAO survey(1) in 1980 showed that 1060 million people in 53 countries were overcutting and destroying their forest resources for fuel, while 100 million people in 27 countries were unable to meet their minimum energy requirements even by overexploiting their forests (Table 2). Dietary habits are being altered to reduce the need for cooking rice and sorghum, which requires much fuel.

The rates of population growth and of forest depletion in these countries suggest that the shortages will get rapidly worse. As wood becomes more scarce, crop residues and dried dung are increasingly used for fuel, thus robbing soils of organic matter and nutrients critically needed for crop production. At present,

TABLE 2
Tropical populations suffering from shortages of fuelwood

	Persons affected (millions)					
	Fuelwood deficits			Acute shortages		
	rural	urban	total	rural	urban	total
Asia and Pacific	710	122	832	29	2	31
Sub-Saharan Africa	131	15	146	49	6	55
Latin America	143	58	201	18	8	26

Source: FAO. 1983. *Fuelwood supplies in developing countries*. Rome.

about 2000 million people depend upon fuelwood, crop residues, and dung for their daily domestic energy needs. By the end of the century, some 2400 million people (nearly half of the world's present population) appear likely to be suffering from acute shortages of domestic fuelwood.(8)

The opportunity for watershed improvements

The fuelwood crisis presents an urgent challenge to those actively concerned with forestry and agriculture in the tropics. The scale of the problem is so great that it can be met only through allocation of very substantial resources by tropical governments and by authorities for overseas aid and technical assistance. Above all, the governments of the fuel-deficient countries must recognize the problem and become politically committed to large and urgent planting programs.

The need for massive planting of trees offers watershed managers a unique opportunity because trees are often the ecologically optimum land use for stabilizing upland watersheds. Where steep upland areas of destroyed forest are now serving only as poor and unproductive livestock ranges with heavy runoff and erosion, the watershed manger has a vital interest in their restoration to more ecologically effective production in which the hillside communities can share.

The basic first step, which many developing countries have yet to take, is to convince both villagers and legislators that the days of the traditional gathering of the nation's fuelwood as a free good are over. Fuelwood must now be planted, tended, and marketed as a crop. Fortunately, during the past half century, demand for industrial timber and pulpwood led to selection and some breeding of fast-growing tree species, particularly among Australia's wealth of drought-resistant eucalypts and acacias. Uniform commercial stands grow more than five times as fast as natural forest and are ready for harvest as pulpwood in 5 to 10 years or as saw logs in 20 to 25 years. They also provide fuelwood as useful residues. Where all the debris is carefully collected, more than 50 percent of the biomass may be available for fuel.

Inadequate investment in reforestation

The 1980 FAO survey of 76 countries showed that a total of 11.5 million hectares of trees had been planted, with an annual rate of planting of 1.1 million hectares.(1) Because the annual

Despite a nutrient-starved agriculture, in India large amounts of dung are gathered and transported to urban areas for use as fuel.

felling of natural forests exceeds 11 million hectares, of which area less than one-half survives to regrow, the replanting rate is disturbingly inadequate. The fuelwood problem is not yet being effectively addressed as an approaching crisis in the resources of the tropical world.

Recognition of the situation grew during the 1980s. The World Bank allocated U.S. $500 million to reforestation programs from 1982 to 1984. The World Resources Institute, established in 1984 to promote action on global resource problems, chose as a first major task the need to arrest the current rate of destruction of tropical forests. A task force convened by WRI, the World Bank, and the United Nations Development Programme set out the scale on which planting for fuelwood, forage, pulp, and timber needs to be undertaken. It recommended investment of some $800 million a year for 5 years to prevent a world problem from becoming a catastrophe. A worldwide increase in investment for tree planting appears to be unavoidable; in this investment, improved watershed management is recommended as a major consideration.(8)

Planting trees outside the forest estate

Since the 1930s, the forest departments of several tropical countries have developed efficient plantations for saw timber, using fast-growing tropical pines and other softwoods and slow-growing hardwoods such as teak. These plantations replaced harvested forests in the reserved areas known as "forest estate." Much of the industrial demand for pulp and fuelwood was met by private plantings made by paper mills and by factories processing tea and tobacco. Fuelwood was also grown for the railways. In East Africa, railway plantations of more than 20,000 hectares of eucalyptus and wattle (*Acacia mearnsii*) were supplying fuel when I traveled in a mainline train in 1945, hauled upcountry from Mombasa by a wood-burning Garrett locomotive. Although an era of cheap fuel oil eliminated this use of local resources, the subsequent tripling of oil prices again focused interest on such biomass plantations for large-scale production of fuelwood in countries where rapid population growth favors labor-intensive technologies. In Kenya, the Forest Department has continued to manage the former railway fuelwood plantations to supply population centers.

Similarly in Brazil in the early 1900s, large plantings of eucalypts were made to supply the railway locomotives. These plantations have since been greatly expanded in the state of Minas Gerais to supply the steel industry with charcoal for smelting. About 2 million hectares of eucalyptus plantations now yield 3 million tons of charcoal annually, so that the industry is self-supporting in fuel. Plantations are expanding by about 100,000 hectares annually, and the total area is expected to reach 4 million hectares by the year 2000. The main species is *Eucalyptus citriodora.* Much of the plantings are in areas where natural forest has been destroyed, reduced to useless scrub, and abandoned.(3)

Addis Ababa provides an outstanding example of a successful response of a tropical city to a growing fuel crisis. The planting of *Eucalyptus globulis* and *E. camaldulensis* in deep volcanic soil in and around the suburbs of the city began about 1910. At an altitude of 2000 meters and with 1500 millimeters of annual rainfall, the trees grew well in spite of the crude methods used for planting. By 1920, they were a dominant feature of the city. Now, the city stands in a wide arc of continuous eucalyptus plantings covering some 13,500 hectares. All are planted, harvested, and coppiced by small-scale farmers, most of whom have less than 1 hectare of trees. Yet in spite of this long-continued and active demonstration of the trees' value, few farmers elsewhere on the high plateau follow this example. The photo of the trees surrounding an Ethiopian religious shrine in Chapter 3 demonstrates that the conditions are favorable, but the traditions are adverse.

In most developing countries, however, such traditional small-scale industries as brick making and lime burning have not progressed beyond reliance on freely gathered fuelwood and are a serious cause of forest depletion. Fuel for these industries must, in the future, be supplied by plantations because the traditional sources are dwindling rapidly, particularly in India.

Selection and breeding for fuelwood and fodder

The opportunities for profitable development of forest plantations are immense. Gains in performance by selection and breeding of fast-growing hardwoods such as eucalypts and fast-growing softwoods such as the tropical pines have been so positive that comparisons of improved strains often outweigh the differences among species. The provenance, or origin, of the seed is therefore critical in plantation development. A major responsibility of forestry departments is now that of securing the provision of selected seed either by direct production or by supervising commercial production. Tree seed production is a highly technical field in which professional forestry guidance is important. Past generations of foresters have been concerned mainly with species and strains selected for saw timber. The expanding modern requirement is for fuel and fodder in addition to timber; thus in-service training to update forestry officers in the silviculture of many recently studied species is important for solving the tropical fuelwood crisis.

An example of remarkable success with the development of a fuelwood species is the work in both Hawaii and the Philippines on a Mexican species, *Leucaena leucocephala.* Giant strains have been produced whose growth rate is more than 10 times that of the original wild species, which botanists formerly named *L. glauca.* (Many foresters and some botanists believe that the giant ipil ipil, as it is called in the Pacific, should be considered a subspecies.)

The original *L. glauca* was so useful as to have been adopted throughout most of the world's tropical lowlands. The giant form grows with astonishing rapidity and has given the highest annual increment of wood production ever recorded, 100 m^3/ha; its usual rate of production is from 30 to 40 m^3/ha. It is an effective nitrogen fixer and thus it enriches the soil. The foliage is high in protein and is a nutritious and palatable fodder, but it should be fed in mixture with other fodders because it contains mimosine, an alkaloid that can produce toxicity if undiluted.(5)

The range of tropical tree species found suitable for rapid biomass production has increased greatly in recent years. Fuelwood plantations can use a wide variety of fast-growing species that would be unsuitable for timber or pulp production because of the quality of the wood or because their stem shape or size prevents the mechanical handling necessary for industrial crops. For steep upland watersheds, tree fodder planted at the lower edge of the forest and in the farmland is, with the cooperation of the farmers, an important means of keeping cattle out of the hillside forest.

Table 3 summarizes the recorded attributes of some species that are likely to be of most use in watershed management in combining fuel and fodder. Some also provide timber, pulp, fruits, or gums. There are many other usable species. Exploration of the wealth of tropical flora for species meeting these objectives has been intensified recently, with some striking results, as from leucaena and calliandra. It is probable that intensive selection among the acacias for performance in dry climate would lead to the development of high productivity.

Planting of species suitable for fuelwood (Table 3) is an important means of overcoming three widespread problems of management in upland watersheds: imperata grasslands in humid areas, bare hills in areas with long dry seasons, and manmade desert conditions.

Imperata grasslands

In regions of high rainfall in Africa, Asia, and the Pacific, the coarse grass *Imperata cylindrica* (lalang) invades areas cleared from forest. This grass, which has little value as pasture, suppresses other growth and presents an annual fire hazard. In Southeast Asia, the area of these "green deserts" has been estimated at 40 million hectares with 16 million hectares reported in Indonesia and 6 million hectares in the Philippines.(7) Although imperata is difficult to control, it is intolerant of shade. At altitudes below 1000 meters, plantings of fast-growing fuelwood species such as *Leucaena leucocephala* and *Gliricidia maculata* can shade out imperata. In the Philippines, a slow but successful establishment was achieved by broadcasting leucaena seeds into the grass; for intensive biomass plantations, however, furrows are plowed along the contour and seeds are scattered into the open furrow.(5) The heavy canopy and dense root mat of *L. leucocephala* and *G. maculata* provide ample soil protection. Because they shade the ground heavily, there is no undergrowth to burn, and both species make effective firebreaks for plantings of more vulnerable species, such as eucalypts.

TABLE 3

Characteristics of species[a] reported successful for fuelwood and fodder

Annual yields (m^3/ha)	Coppice growth	Nitrogen fixation[b]	Fodder quality	Rainfall range, (mm)	Frost tolerance	Soil conditions	Products, uses, notes
				TROPICAL HIGHLANDS			
Acacia mearnsii (A. mollisima, A. decurrens), black wattle							
10-25	Poor	+	Unpalat.	700-1500	Moderate	Not alkaline	Fuel and tanbark
Alnus acuminata (A. jorullensis)							
10-15	Good	+	Poor	1000-3000	Slight	Adaptable	Timber, pulp, erosion control
Alnus nepalensis Nepalese alder							
10-20	Good	+	Poor	500-3000	Moderate	Needs moist sites	Fuel, fodder, erosion control
Artocarpus lakoocha							
Low	None	None	Excellent	1000-3000	Poor	Adaptable	Fodder, fuel
Eucalyptus globulus							
10-30	Good	None	Unpalat.	800-2000	Slight	Deep soils	Poles, pulp
Eucalyptus grandis (E. saligna)							
20-40	Good	None	Unpalat.	1000-2500	Slight	Deep soils	Timber, pulp
Grevillea robusta silk oak							
10-15	None	None	Unpalat.	500-2500	Slight	Deep soils	Shade, timber
Terminalia tomentosa							
Moderate	None	None	Good	1000-2500	Poor	Adaptable	Fodder, fuel

			LOWLAND HUMID TROPICS				
Acacia auriculiformis							
10-20	Poor	+	Unpalat.	1500-2000	None	Adaptable acid or calcareous	Fuel, pulp, erosion control
Calliandra calothyrus							
35-65	Good	+	Edible	1000-3000	None	Adaptable	Fuel, erosion control
Cassia siamea							
15	Good	+	Good but toxic to monogastrics	700-1500	None	Tolerates poor but deep and well-drained acid or calcareous	Fuel, poles
Casuarina equisetifolia							
7.5-25	None	+	Edible	300-3000	None	Salinity tolerated	Fuel, poles, erosion control, tanbark
Derris indica							
Moderate	Good	+	Good	500-2500	Slight	Tolerant of high salinity	Fodder, oilseeds, fuel
Dalbergia sissoo	sissoo						
Moderate	Good	+	Good	700-2000	Good	Tolerant of drought and salinity	Dune fixation, shade timber
Eucalyptus globulis							
10-30	Good	None	Unpalat.	800-2000	Slight	Deep soils	Poles, pulp, selected seed sources
Eucalpytus grandis (E. saligna)							
20-40	Good	None	Unpalat.	1000-2500	Slight	Deep soils	Timber, pulp, poles, selected seed sources
Glyricidia maculata (G. sepium)							
Moderate	Good	+	Good	1500-2500	None	Acid or limestone	Terrace banks, fences fire breaks (seeds toxic)

(continued)

TABLE 3 (continued)

Annual yields (m^3/ha)	Coppice growth	Nitrogen fixation[b]	Fodder quality	Rainfall range, (mm)	Frost tolerance	Soil conditions	Products, uses, notes
Gmelina arborea							
20-35	Good	None	Good	750-4500	None	Fertile soils acid or calcareous	Timber, pulp, fuel
Leucaena leucocephala (L. glauca) ipil-ipil							
30-40	Good	+	Good for ruminants	600-1700	None	Neutral or alkaline only	Timber, fuel, fodder
Sesbania grandiflora							
20-25	Good	+ +	Excellent	1000-3000	None	Widely adaptable	Fuel, fodder, reforestation, pulp, gum, food
Syzygium cumini Java plum							
Moderate	Good	None	Fruits only	1500-10000	None	Widely adaptable	Fuel, fodder, fruits, tanbark
Terminalia catappa							
22-36	Poor	None	No information	1000-3000	None	Adaptable coastal sands to limestones	Fuel, fruit, dune control, timber
ARID AND SEMIARID TROPICS							
Acacia albida apple-ring acacia							
–	Good	+	Excellent	650-1000	Slight	Good farm soils and watercourses	Dry season fodder, 600 kg/ha of edible pods
Acacia cyclops							
1-3	None	Some	Poor	200-800	Slight	Dry sands, coastal dunes, limestones	Erosion control, salt tolerance
Acacia nilotica (A. arabica var. *indica)*							
1-5	Poor	+ +	Excellent	300-1000 or irrigated	Slight	Alluvium, but very adaptable	Fuel, gum, fodder, tanbark

Acacia saligna (A. cyanophylla)							
1.5-10	Good	+ +	Good	250-1000	None	Dry, saline, wet, acid, or alkaline soils tolerated	Erosion control, sand-dune control, fodder, fuel
Acacia senegal (A. verek) gum arabic							
5	Good	+	Excellent	200-800	Slight	Sands or well-drained clays	Gum, fodder, food, fuel, erosion control
Acacia seyal (A. fistula)							
low	Slow	+	Excellent	350-1000	None	Widely adaptable	Fodder, gum, timber
Acacia tortilis (sub-species *raddiana, spirocarpa, heterocantha*)							
5	Good	+	Excellent	100-1000	Slight	Alkaline and shallow soils tolerated	Fuel and fodder, sand-dune fixation
Albizzia lebbek (Mimosa sirissa)							
5	Moderate	+	Good	500-2000	Slight	Alkaline, dry	Fodder, fuel, timber
Azadirachta indica (Melia indica, M. azadirachta) neem							
10-20	Good	None	Poor	450-1100	None	Widely adaptable to nonsaline soils	Timber, oil seeds, pesticide, fuel
Casuarina equisetifolia							
7-20	None	+	Poor	200-500	None	Salt-tolerant bare sands, high temperature	Poles, pulp, fuel shelterbelts, tanbark

(continued)

TABLE 3 (continued)

Annual yields (m^3/ha)	Coppice growth	Nitrogen fixation[b]	Fodder quality	Rainfall range, (mm)	Frost tolerance	Soil conditions	Products, uses, notes
Eucalyptus camaldulensis (E. rostratum) Red River gum							
20-30	Good	None	Unpalat.	200-1250	Moderate	Adaptable by provenance selection and breeding	Fuel, timber, shelterbelts; selected seed sources
Eucalyptus citriodora Lemon-scented gum							
15	Good	None	Unpalat.	600-1500	Moderate	Adaptable to poor soils	Fuel, saw-timber, citronella oil, honey
Pinus halepensis Aleppo pine							
3-12	None	None	Unpalat.	250-800	Good	Mediterranean and dry, subtropics, poor limestones, and heavy clay	Timber, resin, erosion control
Prosopis cineraria							
3+	Good	+	Excellent	200-500	Moderate	Poor saline and alkaline soils, black cracking clays	Fuel and fodder, planted in crops, erosion control, dune fixation
Prosopis juliflora Mesquite							
5-6	Good	+	Excellent	150-750	Slight	Widely adaptable (can be a serious weed in moist sites)	Fuel, timber, fodder, honey, dune fixation
Zizyphus mauritiana Indian jujube							
-	Good	None	Good	300-2000	Moderate	Widely adaptable	Fruit, timber, fodder, fuel, silkworm breeding

[a]Names in parentheses are synonyms. [b]Few quantitative studies of nitrogen fixation in tropical trees have been made. + = evidence of nitrogen fixation; + + = evidence of high rates of nitrogen fixation, such as ample production of leaf proteins.

Sources: National Academy of Sciences. *Firewood crops*. Washington D.C.: National Academy Press, 1980; K.K. Panday. *Fodder trees and tree fodder in Nepal*. Birmensdorf: Swiss Federal Institute of Forestry Research, 1982. And personal observation.

Bare hills

The second general problem arises in climates with long dry seasons. Hills that have been stripped of forests and of much of their topsoil provide such a harsh environment in the dry season that reforestation with timber species often fails. This is particularly the case where old-fashioned foresters insist on a hot dry-season burning to destroy all rival vegetation before planting, a practice that has left many "bald-headed hills" in central China. Some hardy fuelwood species listed in Table 3 can survive and grow in such harsh environments. They will not produce high yields of fuelwood on eroded subsoils, but their root networks stabilize the hillsides and begin the long process of recovering fertility.

In areas of the tropics with more than 700 millimeters of annual rainfall, at altitudes up to 1500 meters, both *Acacia auriculiformis* and *Cassia siamea* have been used to stabilize poor, eroded soils. In dry conditions, an exceptionally hardy shrub or small tree is *Acacia saligna,* which survives on a wide range of

Hills stripped by fuel gatherers in central China have become wastelands that are neither grazed nor cropped. Deep silt traps guard the low-lying paddy fields from sheet erosion from the hillsides.

arid or calcarious soils in hostile conditions. For dry limestone hills in the subtropics, the Mediterranean Aleppo pine *(Pinus halepensis)* will grow on shallow soils over limestone with as little as 250 millimeters of annual rainfall.

Sand dunes and saline deserts

For planting in sand dunes and saline deserts, the acacia and prosopis species given in Table 3 are the most widely used. In addition, two extremely hardy tree species that can survive extremes of heat, cold, and salinity in the deserts of Asia, from western China to the Mediterranean, are the black saksaul (*Haloxylon aphyllum*) and the white saksaul (*H. persicum*). Both will survive saline groundwater. The white saksaul is more drought tolerant (it will grow in areas with 100 to 200 millimeters of annual rainfall) than the black species but less cold tolerant. Both are valuable for stabilizing sand dunes and for providing fodder and fuel. *Prosopis tamarugo* from the north of Chile is a tropical species that shows extreme salt tolerance. In the Atacoma Desert, it is planted through a surface crust of salt and survives.

For arid zone reclamation and dune stabilization in India, Gosh(4) reports successful planting in low rainfall zones with the following:

- Annual rainfall: 150–300 mm
  Acacia senegal
  Acacia tortilis
  Prosopis chilensis
- Annual rainfall: 300–400 mm
  Acacia senegal
  Acacia tortilis
  Prosopis cineraria
  Prosopis chilensis
- Annual rainfall: Over 400 mm
  Acacia senegal
  Acacia nilotica
  Acacia tortilis
  Ailanthus excelsa

Albizzia lebbek
Azadirachta indica
Parkinsonia aculeata
Prosopis chilensis
Prosopis cineraria

References

1. Food and Agriculture Organization. 1983. *Fuelwood supplies in developing countries.* Rome.
2. Food and Agriculture Organization and United Nations Environment Program. 1982. *Tropical forest resources.* Forestry Paper no. 30. Rome.
3. Gallegos, C.M.; Davey, C.B.; Kellison, R.L.; Sanchez, P.A.; and Zobel, B.J. 1984. *Technologies for reforestation of degraded lands in the tropics.* Washington D.C.: U.S. Congress, Office of Technology Assessment.
4. Gosh, R.C. 1982. *Afforestation and management of tropical wastelands in India.* Washington D.C.: U.S. Congress, Office of Technology Assessment.
5. National Academy of Sciences. 1977. *Leucaena: Promising forage and tree crop for the tropics.* Washington D.C.: National Academy Press
6. Revelle, Roger. 1980. Energy dilemmas in Asia. *Science* 209:164–173.
7. U.S. Congress, Office of Technology Assessment. 1984. *Technologies to sustain tropical forest resources.* Washington D.C.
8. World Resources Institute. 1985. *Tropical forests: A call for action.* Washington D.C.

6
The use of trees in tropical agriculture

A major opportunity in watershed management

In many developing countries, a political recognition is dawning that the rapid growth of populations in the past three decades has eliminated the validity of traditional rights to fuelwood and fodder as free goods. New community rules must now be applied to secure the planting and management of fuelwood and fodder as crops.

Tree foliage provides valuable livestock fodder in all climates from humid forests to dry rangelands. Moreover, fast-growing fuelwood trees can be profitable cash crops when they are produced within reach of rapidly enlarging urban markets. The watershed manager therefore has a compelling interest in promoting the planting of trees on farmlands in order to reduce overgrazing and overcutting of steep woodlands. In the longer term, trees planted on farmland will protect soil against wind and water damage, making an important physical contribution to watershed stability.

Agroforestry: The critical effects of competition for water

Agroforestry is a general term for the planting of trees on farmland for fuelwood, poles, shade, shelter, or fruit. It is not usually applied to large-scale plantations of single species such as tea or rubber.

Although people have planted fruit and shade trees around

their homesteads from earliest times, today's improved technologies offer great opportunities to the small-scale farmer. The need to bring modern tree-planting practices to farmland in deteriorating watersheds is urgent. The term *agroforestry* has caused some confusion because writers have tended to apply it both to new methods and to historical practices throughout the whole range of climates from humid to arid. Because the technologies of agroforestry include components of forestry, crop husbandry, and animal husbandry, together with the supporting services meteorology and hydrology, it is not surprising that there has been little organized collection of data. The subject has been obscured in recent years by a surge of semipopular writing that is stronger in enthusiasms than in facts, so that, for the watershed manager, the present state of the literature is confusing. The International Council for Research in Agroforestry, an information and research-promotion center established in Nairobi in 1978, has produced useful surveys and has organized many comparative trials. Only a great deal more measurement and scientific standards of reporting will bring order to this complex subject.

Most of the confusion in the literature arises because critical aspects of water relations are ignored. Rainfall amounts are not specified, and sometimes combinations of trees and crops are described without mentioning that they will succeed only if irrigated. As explained in Chapter 3, trees use more water than do crops of lesser stature for three important reasons:

- The greater foliage area of trees intercepts more rainfall; the film of water on the foliage is subsequently evaporated at free-water surface rates.
- The greater height of trees raises the turbulence of windflow and thus increases the transport of water vapor.
- The greater root range of trees enables them to draw on more of the soil moisture (and sometimes on groundwater) during the dry season.

When interplanted with crops, trees therefore compete for moisture. Competition may be of little consequence under a rainfall of 1500 to 3000 millimeters per year and may not matter at all if ample irrigation is available. But moisture competition is critically important for rainfed crops; the severe dry spells that occur within the erratic rainy seasons are a characteristic of the Intertropical Convergence Zone.

Because watershed management authorities should support tree planting by farmers, it is important that these water relationships be understood. Some useful examples of successful technologies follow. Many more can be found among the endless combinations of prolific tropical flora and the variety of customs among tropical communities.

Tree crops in the humid tropics

Tree gardens and soil stability

Among the many traditional patterns of subsistence farming, the one most favorable to watershed stability is the tropical tree garden (see Chapter 2). In tree gardens, cultivators skillfully replace a multistoried rainforest with a multistoried array of tree crops that provide food and cash. In the Kandy area of Sri Lanka, in the islands of Java and Bali in Indonesia, and in Sabah, I have observed that this intensive land use generates a relatively better standard of living, even from family holdings of less than 1 hectare. A detailed study of tree gardens has been made by Bompard et al.(2) However, when rapid population growth forces new farmers out onto steep, roadless hills or into areas of lower rainfall, the close relationship between poverty and soil erosion is demonstrated. In Java, a densely settled island that experienced a 66 percent increase in population between 1961 and 1980, subsistence farming has spread onto thinner soils and steeper slopes, which were cleared of forest and planted to cassava or soybeans. This has produced a series of watershed problems that have increased sediment flow from misused hills, thus threatening lowland irrigation systems. Improvements have been demonstrated by encouraging the planting of trees and by applying well-established principles of soil conservation (Chapter 8). Improved methods are being extended by government departments with the support of overseas aid teams.(1)

Production of commodity tree crops

The large-scale estate production of palm oil, rubber, tea, coffee, and cocoa has traditionally been based on competently engineered roads and drainage. Recent "outgrower" plantings on small farms are less well organized. I have seen plantings of rubber in both Thailand and Malaysia that ignored hill contours, and consequently, despite adequate natural ground cover, the footpaths of the rubber tappers caused severe erosion in unstable soils. In addition, obsolete types of silt traps—deep trenches dug across steep slopes with excavated soil piled on the downhill edges—undoubtedly increased soil movement downslope rather than preventing it. In recent years, many of the less successful rubber plantings have been cleared and replaced by oil palms. The heavy fruit bunches are conveyed to the factory in 5-ton trucks so that the need to redevelop the plantations with roads, terraces, and hillside drains has led to some spectacular designs of contour planting visible for many kilometers.

A dense ground cover of weeds protects the soil but involves labor costs for slashing by hand. Replacement by nitrogen-fixing leguminous cover crops reduces the need for fertilizers and the costs of control. It is now standard practice in new plantings of oil palm, rubber, and coconuts and has been adopted in many older plantations. Work at the Rubber Research Institute of Malaysia has shown that during the 35-year life of a rubber plantation, legumes can contribute about one third of the 2 tons of nitrogen required per hectare, which is either immobilized in the trees or harvested as latex. The rest must be supplied by fertilizers if latex yields are to be maintained. The University of Malaysia has shown that oil palm is able to derive full nitrogen requirements from legumes, giving no additional response to nitrogen fertilizer. Such cover crops are therefore very important for the small farmer.

Successful livestock production under these tree crops has been developed in Malaysia. The Rubber Research Institute has shown that a natural weed cover can support two sheep per hectare whereas a full cover of legumes can carry 12 sheep per hectare. Under the high standards maintained by these plantation industries, the watershed manager has few problems of soil erosion. Legume species that have proved sufficiently shade tolerant to thrive in these conditions are *Pueraria phaseoloides, Calopogonium caeruleum,* and *Centrosema pubescens.* It is important for watershed management that these improvements in technology should be adopted by the small-farm outgrowers

At the drier limits of the areas in which plantation crops are grown, such vigorous cover crops may compete for water during the short dry seasons. Pereira and Hosegood(14) reported that kudzu in clove plantations in Zanzibar increased drying in the first meter depth of soil.

Trees for shade and shelter

In the wild, tea, coffee, and cocoa are part of the understory vegetation, growing to about 6 meters in height. The early commercial plantings of these beverage crops were therefore made in heavily thinned forest. Colombia still grows some of the world's highest quality coffee under forest shade. Cocoa in West Africa is similarly grown under shade. Interplanting shade trees of uniform height and spacing was the next development. For shading tea in East Africa, *Grevillia robusta* has been used almost exclusively; both neem *(Azadirachta indica)* and *G. robusta* are widely used in India. Neem has the remarkable property of repelling insects.

Shading, however, reduces the yield of cocoa. Research at the West African Cocoa Research Institute in Ghana has shown that the main benefit of shade on cocoa was control of the capsid bug.(4) In a trial planted in 1959, in which cocoa was grown with and without shading by *Terminalia ivorensis* and with pests controlled by insecticides, 6-year mean yields (1964–1969) were 85 kg/ha of dry beans from the shaded trees and 172 kg/ha from the unshaded trees. Fertilizer responses were 4 percent under shade and 33 percent without shade. However, newly planted cocoa benefits from light shade during its early growth. The recommended practice is to plant fast-growing leguminous trees such as *Gliricidia maculata* in advance of the cocoa seedlings and remove them when the cocoa has formed a closed canopy. In the Commonwealth Development Corporation cocoa plantations in Sabah, a double row of gliricidia is left around each plot of about 5 hectares, forming shelterbelt boxes to provide protection from storms.

In tea plantations in Kenya, unshaded rows yielded more and better quality tea than shaded rows.

In Kenya, shade-tree research in tea gave similar results. The East African Agriculture and Forestry Organization found that tea yields are responsive to total daily radiation at the plucking surface. Highly significant results were obtained at Limuru under 1300 millimeters of annual rainfall and at Kericho under 1800 millimeters. Both sites are near the equator at about 2200 meters altitude. Compared with tea grown under grevillia trees at 35 to 40 per hectare, tea grown without shade yielded 10 percent more, and when nitrogenous fertilizer was applied, 25 percent more.(12)

Questions of the effect of shade on tea quality were settled by plucking separately the more heavily shaded rows under the trees and the less-shaded rows midway between them. Bulked over large areas, the two pluckings were separately processed and sold on the London market. The less-shaded tea fetched a higher price.

Because the grevillia trees have to be felled after 10 years to minimize the physical damage to the tea bushes, the total cost of planting and tending them is substantial. As a result of the research, the use of shade trees in tea plantations has declined in East Africa, in Sri Lanka, and, to a lesser extent, in India. But because hail can cause severe damage to tea crops, protection by shade trees is essential where hail is a hazard.

Around the edges of tea plantations, the planting of shelter trees as windbreaks has increased. Research has shown that windbreaks have a beneficial effect, provided that rainfall or irrigation is adequate, as they are in most of the world's tea-growing areas. The research also has revealed an unexpected effect for tea grown near the limits of its environment in long dry seasons without irrigation. Hot winds cause tea bushes to close their stomata, while sheltered bushes continue to transpire and eventually suffer more severe drought stress.(3)

Grevillia robusta is used for shade and windbreaks because its root systems do not compete excessively with the tea. Both *G. robusta* and tea root to more than 4 meters depth, where the soil allows, and in drought conditions tea bushes near the trees show no visible stress. In contrast, shelterbelts of eucalyptus species, which have a horizontal root range of 5 meters or more, have been observed to have adverse effects on neighboring tea bushes, except in areas of high rainfall. Shelterbelts are important along coastlines in the humid tropics, as in South China and West Africa. Large-scale plantings of shelterbelts have been made to stabilize sand dunes and to protect agriculture from high winds.

An impressive example is the "great green wall" along the tropical south coast of China. The belt is 3 kilometers wide and some 3000 kilometers long. Planting, mainly of *Casuarina equisetifolia,* began in 1954 and continues, with more than 1 million hectares now completed. Sand-dune fixation and protection from sandstorms and typhoons are important objectives. The shelterbelts also supply fuelwood and construction timber.(17) The area is greater than that of the Great Plains Shelterbelt in the U.S.A., which was planted from 1935 to 1942 from the Canadian border to Texas to check wind erosion during the Dust Bowl period.

Protection is also important for individual farms. Two rows of trees along the windward boundary of a farm can improve crop yields as well as provide firewood and fodder. The physical effects of shelterbelts are discussed in the section that follows, in the context of drier climates, in which their benefits are greatest.

Intercropping with trees

The main tropical tree commodity crops can be grown in combination, sometimes affording high cash returns per hectare. For example, the tall tree crops can be spaced and oriented to give light for underplanted cocoa. In Sri Lanka, yields from rubber trees spaced at 10×2.5 meters equal those of standard plantings at 7×3.6 meters, but the wider spacing permits rewarding crops of underplanted cocoa in the drier areas of some 1500 millimeters of annual rainfall (in wetter areas, however, this combination fosters fungal diseases). When the rubber is to be felled and replanted, the cocoa is stumped—that is, cut back to ground level to regrow with the new rubber trees. Cocoa in the Philippines is similarly planted under coconuts grown on the heavier soils. Robusta coffee in Sri Lanka is an alternative undercrop in rubber. In Sri Lankan tea plantations, rubber has also been grown as a shade tree (but only at about one-third of the normal planting density for rubber trees).

In Tanzania, the Chagga tribe has traditionally managed Arabica coffee with good yield and quality under banana shade on the slopes of Mount Kilimanjaro. Elephant grass (*Pennisetum purpureum*) grown on the sloping risers of well-built terraces is used as a mulch in the coffee. It is also fed to penned livestock together with banana foliage. This system has long maintained hillside stability under a rainfall of 1500 millimeters in highly erodible volcanic soils derived from ash and tuff. The close relationships between good watershed management and high standards of farming are well illustrated here. With good technical advice, the system flourished for a half century and became a Chagga tribal cooperative. A few recent years of state management have caused partial failure through overcropping and soil erosion, so that a rehabilitation project has been required.

The interplanting of trees with arable crops in the humid tropics occurs mainly during the period of establishment of the tree crops. Food crops are grown during the land preparation, planting, and tending of the young trees. The system, known in

Southeast Asia as taungya and in Africa as shamba planting, has been widely used by tropical foresters in countries where land is in short supply. It is, in effect, the organization of shifting cultivation with priority for trees rather than for food crops. The cropping sequence usually lasts only 2 or 3 years before the new trees shade out the crops of maize, sorghum, millet, cassava, sweet potatoes, and squashes. An exception is in high areas of Pakistan, where the very slow-growing Chir pine (*Pinus roxburghii*) is intercropped for 10 years or more. When the tree canopy closes in 3 or 4 years, the forest topsoil is then still rich enough in accumulated organic matter to maintain stability. Provided that the food-crop rows were planted across the slope, no erosion need occur (as shown by the hydrological watershed experiment in Kenya described in Chapter 3). Taungya crops should not be grown on the stream banks, which should be left under a protective strip of forest. A danger from interplanting is the forestry tradition of planting tree rows downslope rather than on the contour. This practice has advantages for field control of labor, but it can have serious results if intercrops are also planted downslope.

Leguminous shade trees can make important contributions to soil fertility. The high rainfall of the humid tropics rapidly leaches soluble nitrogen compounds from the soil. The temperate-zone practice of growing leguminous field crops and plowing them in can be paralleled on small farms in the tropics by growing leguminous trees and using their foliage as a mulch. The modern strains of *Leucaena leucocephela,* for instance, can fix up to 600 kg/ha of nitrogen annually and can yield 100 t/ha of fresh foliage (8 t/ha dry matter), as compared with 30 to 40 t/ha from unimproved strains.(13) Where rainfall is adequate to permit food crops and leguminous trees to be grown together, nitrogen fixation by the trees can be of direct benefit. In western Nigeria, the International Institute for Tropical Agriculture has experimented with leucaena interplanted with maize in an area of 1500 millimeters annual rainfall. The leucaena is closely planted in rows that are widely spaced and aligned east-west to minimize shading of the maize. The trees are cut back heavily while the crop is growing so that the young maize can be mulched with the nitrogen-rich foliage and green stems. Improved soil stability with bonuses of fuelwood and forage are claimed, without reduction of maize yields.

Tree crops in medium-rainfall areas

Shelterbelts or windbreaks

Tropical lands in the middle range of annual rainfall (600–1000 mm) usually have an open woodland or tall-tree savanna type of ecology. Dry seasons are characteristically long and severe and crops suffer from exposure to hot winds and blown soil; thus the planting of trees as shelterbelts is an important aspects of farming.

The effects of shelterbelts have been closely studied in wind-tunnel and field tests in Europe, the U.S.A., and Japan. The findings can be applied to the tropics because the physical behavior of airflows over and around obstacles is universal. The studies show that the objective is not to plant a thick hedge that makes a solid barrier to the wind. When the wind strikes an impermeable barrier, it is deflected over the top and causes severe eddies on the leeward side that may damage crops. A solid shelterbelt is effective for a distance downwind that is only some ten times the height of the barrier. The sheltered zone may be doubled by planting a windbreak with about 50 percent gaps. This arrangement breaks up the windflow into small eddies, so that the wind speed is reduced by half for a distance of about 20 times the height of the shelterbelt (Fig. 24). Although wind-tunnel studies of holes in flat-plate barriers give 35 percent as the optimum, with trees, in practice, 50 percent gaps are advisable.

The need for windbreaks is evident not only from scientific measurement but from simple observation of the effects of wind on common crops such as maize in the dry season. The outer few rows on the windward side of the crop are often wilted and stunted. The effect is also apparent in tree crops. In Kenya, Arabica coffee in the lower edge of its altitude range has to be protected by windbreaks. Even when irrigated, the outer two rows on the windward edge of the plantations develop thickened yellow leaves and yield poorly unless protected from dry winds.

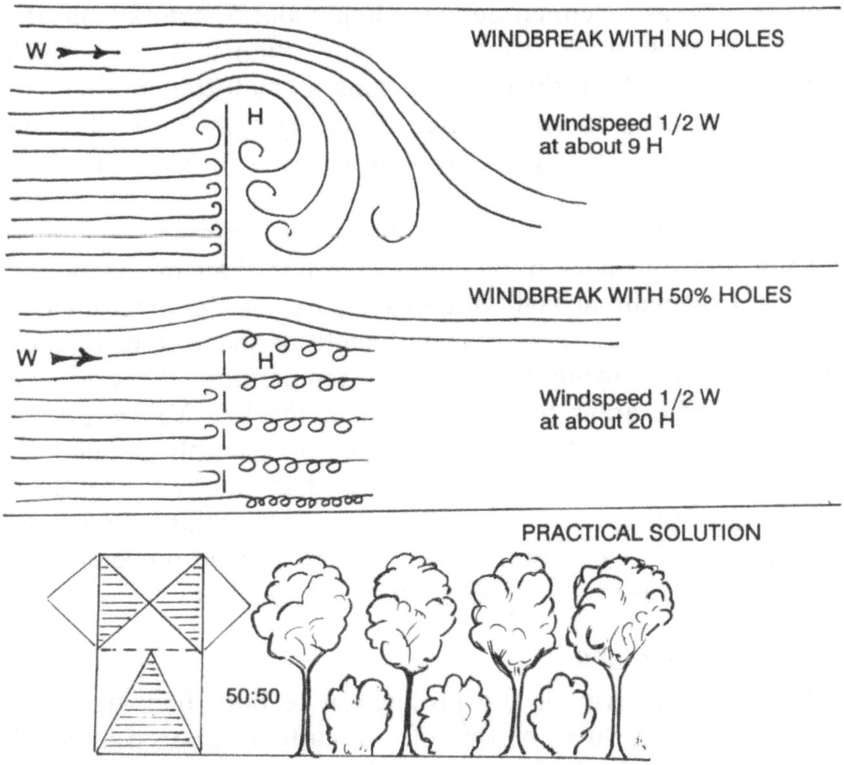

Figure 24. Windflow over shelterbelts. For a shelterbelt of a given height (H), permeability reduces wind speed (W) for a greater distance than a dense shelterbelt of the same height. In practice, the aim should be 50 percent permeability.

Measurements in the Sudan Gezeira(10, 15) of the rates of evaporation at crop-surface level near the windward borders of unsheltered fields of cotton and alfalfa showed severe water losses for the first 50 meters. In the central Great Plains of the U.S.A., Hanks, Allen, and Gardner(8) made an elaborately instrumented study of energy exchange and water loss from a sorghum crop. They found that advected energy (heat input from hot winds) increased crop water losses by 30 percent over the first 40 meters of unsheltered fields.

In the dry season, the roots of the windbreak trees will of course compete with those of the adjacent rows of crop. The commonly grown species of eucalyptus have vigorous surface root mats so that the effects of competition during a dry spell may be seen as a semicircle extending about 5 meters from the trunk. On very small farms, this competition will be of much concern. It can be reduced by digging a small trench about 60 centimeters deep to cut the tree roots between the shelterbelt and the crop (but if dug too near to the trees, this can result in wind throw).

In this middle rainfall range, temperatures vary both seasonally and diurnally over a greater range than in the humid tropics, especially at high altitudes such as on the main plateau of southern Africa. An important benefit of shade trees in such areas is the narrowing of the diurnal temperature range. In the Arabica coffee areas of Kenya, east of the Rift Valley, for instance, unshaded coffee flourishes up to a sharply defined altitude limit of 1750 meters. Above that, the wide diurnal temperature range causes a proliferation of branching, known as "hot and cold" symptoms. This problem can be prevented by shading the coffee with tall *Grevillia robusta* trees.

Trees in grazing land

The ideal pattern for the production of both fuelwood and forage would appear to be that of the scattered-tree savannas of Africa between the Sudan and the Limpopo River in southern Africa, where the trees provide shade and shelter in addition to wood. Similar areas of South America offer parallel opportunities.

An intriguing point for soil and water management is that in spite of the broad surface root mats of the dryland acacias (Fig. 22), the grasses under the trees are usually taller and greener than those in the open areas. The physical benefits of the shade appear to be supplemented by the enrichment of the soil from leaf fall and the dung and urine of livestock that are seeking shade. Similar findings have been reported from the semiarid tropics of Australia. Livestock seeking shade under the tall *Eucalyptus populnea* in Queensland dairy pastures so increase the soil fertility that by broadcasting seed of the deep-rooted *Cenchrus ciliaris*, patches of improved pasture have been developed.(5)

Management, however, presents real difficulties both for the watershed authority and the livestock farmer because the ecological balance in medium rainfall climates is precariously maintained by frequent grass fires. Ecologists believe that in addition to fires ignited by lightning strikes, they have continually been set by hunters, honey gatherers, and herders for many centuries.(16) Over the major savanna areas dominated by acacia species in Africa, fire is essential to suppress the growth of tree seedlings, which otherwise produce dense thickets. Livestock intensify the problem by consuming the edible seedpods and distributing viable seeds in their dung. Furthermore, overgrazing both destroys protection against soil erosion and eliminates fuel for grass fires, so that dense thorn scrub thicket replaces the grassland. Proper grazing management must therefore include closure of areas in succession to permit grass to grow enough for a bush-controlling fire every 4 or 5 years.

Watershed managers must be actively concerned with controlling fires. Much damage to soil protection occurs when grass fires destroy trees and shrubs on steep slopes. Where such slopes are in the care of foresters, firebreaks are maintained, which involve substantial costs. Organized back-burning of protective strips can be effective, but it requires supervision and labor. Only one person with matches is needed to start a fire that can require 20 or more to control it. (The difficulties of controlling fires on communal grazing lands are discussed in the next chapter.)

While the acacia savannas give management problems of thicket formation and fire control, the taller types of open woodland such as the *miombo* of Zambia (in which *Brachystegia* species are dominant) and the *mopane* of Zimbabwe (in which *Colophospermum mopane* is dominant) are rapidly losing their fuelwood trees. The mopane woodlands of Zimbabwe are showing the usual effects of population pressure, and more than half of the communal areas have been cleared of trees, leaving no saplings for regrowth. Here watershed management requires replanting of fuelwood species especially to prevent the loss of protective trees on steep slopes.

As plant breeding and selection have progressed, productive combinations of tree species and grasses, sometimes described as sylvo-pastural systems, are becoming important options for tropical land use. For example, in an experiment involving intensive management of dairy pastures in northeastern Australia, rangeland was plowed and planted to fast-growing strains of poplar trees at 6 × 6 meters. Weeds were controlled for 2 years by growing a ground cover of melons and squashes, and then improved strains of pasture grasses were sown. These steps raised the cattle-carrying capacity in addition to providing a crop of wood. The cattle benefited from the shade and from the palatable forage provided by fallen poplar leaves.(11)

Tree crops in semiarid lands

Accounts of tree growing on small farms in dry country with less than 600 millimeters of annual rainfall need to be read with caution because some popular articles do not specify whether or not the trees are irrigated.

Trees in irrigated fields

Planting of trees along the borders of irrigated fields is traditional in China and is a major component of the 30 million hectares of trees planted since 1949. Along the desert edge near the borders of Mongolia in Gansu province, a shelterbelt 150 kilometers long consists of closely spaced poplars grown in four or five parallel rows about 100 meters apart. Within the rows, the trees are planted 1 to 2 meters apart. Between the rows are continuous paddy fields divided into narrow strips as individual holdings. A special irrigation furrow serves the trees, and there is no visible adverse effect on the rice crops.

Trees interplanted with crops are an important feature of India's major wheat-producing areas, Haryana and Punjab. Under ample irrigation, poplars for pulpwood are a major crop on farms of 50 to 100 hectares. On one particularly well-managed farm I visited in Haryana, widely spaced poplar trees, planted in north-south rows to minimize shade, are underplanted, in a 4-year rotation, with sugarcane for 2 years, followed by wheat and then by potatoes. Catch crops of the legume crotalaria are grown to provide nitrogen to decompose the massive root systems of sugar. On other fields, large mango trees are intercropped with wheat, while melons are grown in the shade under each mango. With ample sunshine and water, the combination of tree crops

In northern China, a shelterbelt of poplars spaced 1 meter apart protects rice paddies. The trees are irrigated from a separate channel.

with arable crops gives maximum productivity. A point of importance for management is that the pulp mills of Haryana and Punjab maintain extension agents who provide seedlings of fast-growing strains of poplars to farmers.

For building poles and fuelwood, eucalypts are also grown among irrigated crops by farmers in India; the Mysore hybrid (a strain of *Eucalyptus tereticornis*) is most used. Studies on interplanting eucalypts with wheat at Dehra Dun by the Soil Conservation Institute of the Indian Council of Agricultural Research showed that wheat yields were reduced 10 to 15 percent. Fuelwood and poles compensate for this reduction, but even when closely planted, eucalypts alone do not meet the costs of irrigation projects.

Trees in rainfed crops

Some farming communities of India and Africa in areas of 400 to 500 millimeters of annual rainfall have lived long enough with scarcity of fuelwood and of fodder to have developed traditions of planting trees among their crops. Unfortunately, they consitute

Eucalypts grown in irrigated wheat reduced grain yields 10 to 15 percent in Indian studies.

a very small fraction of the total population currently at risk from these shortages.

In Rajasthan the traditional tree for the croplands is the thorny legume *Prosopis cinneraria,* which is lopped for forage so that shading of the crops is minimized during the rainy season. The trees are deep-rooted and continue to produce green forage in the dry season. The more hardy *Prosopis juliana* self-seeds in areas too dry or too rugged for cropping. Both species are slow growing. A popular species with farmers is *Acacia nilotica,* an early introduction from Africa. Farmers have selected a strain with upright branches and a narrow growth habit similar to many cypresses, which is recognized as a subspecies: *Acacia nilotica* s.sp *cupressiformis.* Agricultural officers believe that such interplanting of trees among rainfed crops causes uneconomic reduction of yields, and they are currently advising farmers to restrict plantings to windbreaks along the margins of fields.

Farming communities in the Kordofan province of the southern Sudan also have traditionally planted trees among crops. In this area, which has a rainfall regime of 300 to 400 millimeters, the original vegetation was a scattered-tree open savanna of *Acacia senegal.* Gum arabic, a resinous gum exuded when the bark is damaged, has been collected and traded for some 2000 years. Under the traditional agroforestry system, land was cleared for fuelwood, cropped for sorghums and millets for 3 to 6 years, and then protected by the herders to permit regeneration of acacia from the abundant self-sown seed. From an age of 5 years to about 12 years, the trees were tapped for resin. Seedpods were fed to livestock and branches were lopped for fodder and fuel. As their vigor declined, the trees were felled for fuelwood and the land was again sown to food crops.

In the past 20 years, however, the familiar problems of population increase caused a slow decline and eventual breakdown of the traditional farming system in Kordofan. A rapid expansion of mechanized agriculture in the areas to the south (which receive more than 600 millimeters of rainfall) reduced the grazing areas and thus increased the pressure of nomadic livestock. The rapid increase of farming populations reduced the system to continuous cultivation with meager yields. The result has been an increase in desertification. A vigorous international aid scheme funded by the Netherlands and Sudanese governments began in 1981. Under it, the Forest Department, employing villagers, planted 4 million acacia trees in 3 years. This project showed that the problem of desertification could be solved on 15,000 hectares using a team of 6 forest rangers and 32 forest guards. There are, however, half a million hectares requiring such treatment. About 1 percent of the population has been involved. This technical input demonstrates that the poverty of the area and desertification are unnecessary. It also emphasizes that the main problems of organization for improved land management and for the reduction of population growth in the whole region are not yet being addressed with the urgency needed to win the race against future famine.

Social forestry

Tree planting by village communities

The Kordofan example illustrates how unlikely it is that the growing crisis in fuelwood supply in tropical developing countries can be solved while tree planting is done only by government forestry organizations. There is no possibility of training and deploying the vast army of technical staff (and the supporting bureaucracy) that would be required. Nor is this necessary. It is evident that subsistence farmers in all but the most humid tropical lands must adopt fuelwood and forage species as crops. They must learn to plant, tend, harvest, and market surpluses as they already do for food crops. This in no way relieves the forest departments of responsibility for providing technical support on the same basis as the agricultural departments help with crop and livestock husbandry.

The husbandry of tree seedlings is a skill in which villages need guidance from professional foresters. Commercial supplies of tree seeds in developing countries are often poor in germination and in genetic uniformity because they are collected from scrub forest rather than from seed orchards. Forest departments should be involved in supervising the importation and testing of fast-growing species, in organizing the supply of good-quality seed from selected provenances of species suitable to the area, and in advising on establishment of nurseries and methods of planting. Villagers can then solve their own problems of fuelwood, shelterbelts, and stabilization of soil. This has become known as social forestry.

India, which faces fuelwood problems on the largest scale, has some of the best examples of solutions. The Forest Research Institute at Dehra Dun has magnificent buildings designed by Lutyens and a fine tradition of service. It has produced a force of disciplined and devoted foresters who have dominated forestry practice throughout the subcontinent for more than a century. Their objectives were, however, focused firmly on the production of fine timber for world markets. Faced with a fuelwood crisis, the government of India included a major tree-planting program in the 1980–1985 Sixth Plan. As a result of a major change of policy, half of the 4 million hectares of new plantings have been made outside the government forest estate. Some of the plantings are on industrial enterprises (as described in Chapter 4), but most are part of social forestry schemes for rural communities.

An outstanding success has been in Gujarat state, north of Bombay. The program was begun by the Forest Department in the early 1970s. It was expanded in 1979, with World Bank support, to encourage rural communities to raise seedlings for issue to farmers at a rate of 30 million annually. The program was adopted with such enthusiasm that the rate has increased to 200 million per year. Schools and private farms have developed nurseries as supplementary sources of income. The Forest Department contracts with them to produce seedlings at U.S. $2.00 per 100 and distributes the seedlings free to the farmers. Some 800 schools take part.

The farms are small: Four-fifths of the seedlings have gone to farmers with less than 4 hectares, and one-third have gone to farmers with less than 2 hectares. Shade trees and fruit trees have been planted around households and eucalypts around rainfed fields. In a rainfall of about 500 millimeters per year, the most lucrative returns have been from eucalypts grown in or around irrigated cotton fields. The main cash return has been from building poles, but large quantities of tops and branches are used as fuelwood.

A point of practical importance for other countries has been the development of a low-cost method of raising and distributing tree seedlings in Gujarat. Because the conventional practice of producing seedlings in polythene pots or sleeves is costly in both materials and labor, they have been replaced by bamboo baskets about 0.5 meter in diameter and 15 centimeters deep in which about 2000 seedlings can be germinated. They can be carried more easily than the individually potted seedlings, and the seedlings can be pricked out by the farmers into individual containers. This has reduced seedling costs to U.S. $3.00 per 1000.

Community woodlots have had less success in most developing countries, other than South Korea, because they are dependent on the strength of leadership and social cohesion in the villages. They have survived best along roadsides, railway lines, and canal banks where they can be conveniently and economically supervised by forest department staff. Their establishment offers useful employment to landless laborers. Success, however, depends critically on the will of the community to honor agreements about the protection of such woodlots from livestock and from theft. Often villages must post their own guards. Social organizers see village participation as a major advance on the policing of the state forestlands by forest guards, which has never been fully effective against heavy population pressure. Nonetheless, policing of state forests will always be necessary to ward off urban predators with trucks and labor gangs. Rural villagers will not walk long distances to invade forestland if they have their own supplies grown close to home. This change, which the World Bank is supporting in many countries, has important prospects for improvement of watershed management.

Tree crops in water harvesting

An age-old form of community tree planting in watershed development was based on the collection of water running off from hill slopes. Some of the water-harvesting technologies developed in the Mediterranean region are applicable in other arid lands. The trapping of overland runoff from semiarid hill country, to fill small underground storage cisterns or to saturate small plots of deep soil, maintained successive regimes in the steep and rugged dry hills of the Negev Desert from 2000 B.C. to about A.D.640. In this area, the total annual rainfall of 100 to 120 millimeters arrives in 15 to 25 days of a brief winter and is followed by a severe dry summer. The technology reached its peak in the Nabatean culture in the third century A.D. as part of the Byzantine half of the Roman Empire. Prosperous communities of many thousands were maintained by skillful collection, channeling, and storing of the scant and erratic rainfall from a landscape of which 80 percent was barren rocky slopes.

They stored drinking water in underground cisterns that were quarried from soft limestone rock, lined with stone walls, and sealed with plaster made from burnt lime. A modern exploration discovered 40 such cisterns within a radius of 10 kilometers; all were in repairable condition. Modern Israelis revived agriculture in this formidable wilderness from the small kibbutz of Sdeh-Boker, which has subsequently developed into a university campus and should provide a valuable source of drylands information.(9) Some of the original Nabatean runoff farms were renovated and operated successfully.(6) The yields of water were only 15 percent of the seasonal rainfall, although up to 30 percent of single storms were collected. Cropped areas were 0.4 to 4 hectares, fed by 10 to 100 hectares of sloping catchment to support barley and sorghum and several species of tree crops.

Trees have greater ability than cereal crops to survive on such erratic irrigation of pockets of deep alluvial soil. Experiments in the northern Negev under a rainfall regime of 200 to 250 millimeters have maintained almond trees successfully with runoff areas only three times that of the cropped area.(9) Carobs, guavas, figs, and pomegranates are among the crops grown by this system. Where salinity is a problem, date palms, which have a high tolerance for salts, are used.

Harvesting of runoff is traditional on the borders of deserts in Asia, as in Rajasthan in India. In Baluchistan, in Pakistan, collecting dikes are now made by modern earth-moving machinery.(7) The recently developed UNESCO/ICAR Central Arid Zone Research Institute at Jodhpur in Rajasthan, India, is also undertaking studies on water harvesting to maintain fruit trees of which *Zizyphus mauritania* and figs are important examples.

References

1. Barrau, E.M. 1984. *Production constraints and soil erosion in the humid tropics of densely populated Java.* Jakarta: U.S. Agency for International Development.

2. Bompard, J.; Ducatillion, C.; Hecketsweiler P.; and Michon, G. 1980. *A traditional agricultural system: Village, forest and gardens in West Java.* Montpellier, France: Universitáe des Sciences et Techniques du Languedoc (in French).
3. Carr, M.K.V. 1985. Some effects of shelter on the yield and water-use of tea. *Progress in Biometeorology* 2:127–144.
4. Cunningham, R.M.; Smith, R.W.; and Hurd, R.R. 1961. A cocoa shade and manurial experiment at the West African Cocoa Research Institute, Ghana. *Journal of Horticultural Science* 36:116–125.
5. Ebersohn, J.P. 1969. Grazing management of dairy pastures. *Queensland Agricultural Journal* 95:574–582.
6. Evanari, M.; Shanan, L.; Tadmore, N.; and Aharoni, Y. 1961. Ancient agriculture in the Negev. *Science* 133:979–96.
7. French, N., and Hussain, I. 1964. *Water spreading manual.* Range Management Record no 1. Lahore: West Pakistan Range Improvement Scheme.
8. Hanks, R.J.; Allen, L.H.; and Gardner, H.R. 1977. Advection and evapotranspiration of wide-row sorghum in the central Great Plains. *Agronomy Journal* 63:520–527.
9. Hillel, Daniel. 1982. *Negev: Land, water and life in a desert environment.* New York: Praeger.
10. Hudson, J.P. 1963. Variations in evaporation rates in Gezeira cotton fields. *Empire Cotton Growing Review* 40:253–261.
11. Lober, K. 1980. Agroforestry productivity. *Australian Forest Grower.* 3:31.
12. McCulloch, J.S.G.; Pereira, H.C.; Kerfoot, O.; and Goodchild, N.A. 1965. Measurements of shade-tree effects on tea yields. *Agricultural Meteorology* 2:385–399.
13. National Academy of Sciences. 1977. *Leucaena: Promising forage and tree crop for the tropics.* Washington D.C.: National Academy Press.
14. Pereira, H.C., and Hosegood, P.H. 1962. Soil moisture effects of kudzu as a clove-orchard cover crop. *East African Agriculture and Forestry Journal* 27:225–229.
15. Rijks, D.A. 1971. Water use by irrigated cotton in the Sudan (3). Bowen ratios and advective energy. *Journal of Applied Ecology* 8:643-663.
16. UNESCO/UNDP/FAO. 1979. *Tropical grazing land ecosystems.* Paris: Natural Resources Research Department, UNESCO.
17. Zhu Zhir Song. 1981. China's great green wall. *American Forests* 87:58–59.

7
The role of livestock in watershed management

Profitable farming or resource destruction

Well-managed livestock pose few problems of watershed management in the more technically developed economies. For the catchment areas of reservoirs, carefully stocked and closely grazed sheep pastures are a very satisfactory land use. They provide both maximum soil protection and minimum transpiration losses.(19) Under the more intensive stocking that occurs in dairies and feedlots, an important concern is that concentrations of dung and urine can jeopardize the quality of both groundwater and surface water. Methods of treatment and disposal are well established but expensive, so that their application must be enforced.

In contrast, ill-managed livestock are one of the major sources of watershed damage in tropical developing countries. This damage occurs throughout the whole range of ecological conditions from steep forested mountain slopes under high rainfall, as in the Himalayas, through the wet lowland forests of the Amazon to the dry African savannas of the Sahel. In all these contrasting conditions, the effects of livestock serve to increase the vulnerability of watersheds to the high intensities and erratic patterns of tropical rainfall under the influence of the Intertropical Convergence Zone (ITCZ).

The importance of the transfer of plant nutrients

In spite of much unnecessary soil erosion caused by lack of grazing management, livestock have an essential role in many forms of subsistence agriculture. They not only provide meat, milk, and other products, including draft power, but they harvest plant nutrients from herbage, concentrating nitrogen, phosphorus, and other essential elements in dung and urine for the nutrition of crops. Where human population growth has left no land available for the shifting bush fallows that formerly restored some soil fertility, continuous cropping has so depleted the soil of nutrients that crop yields have dwindled. Until the economy and the road network are able to deliver fertilizers, subsistence farmers on overcrowded lands cannot even recover a yield equal to their seeding rates unless they apply manure. This lack of plant nutrients has been the basic cause of the recent famines in Ethiopia and the Sahelian countries, which are the culmination of a long period of population growth and soil exhaustion. In the late 1960s, in an eroded area of Ethiopia, I saw wheat and barley crops 20 centimeters high that were ripening with only one or two shrunken grains in each head.(18) The rains had been adequate, but the plant nutrients were exhausted. Drought in 1984 finally brought collapse.

Livestock, however, can only harvest the nutrients if the herbage is first permitted to grow. Competitive overstocking of common land is so severe in many tropical countries that the soil is stripped bare. Shrubs that could provide edible foliage are removed for domestic fuel. The result is barren wasteland. In India, there are 13 million hectares of such wasteland, and one-third of the 20 million hectares of gazetted government forest estate requires active measures of soil conservation as the result of overgrazing. That area is in addition to the 58 percent of India's cropland that is in need of soil conservation treatment.(26)

Livestock in steep hill country

Along the vast ranges of the Himalayas from Pakistan to Sikkim, watersheds vital to the crowded Indo-Gangetic Plains are threatened by increases in populations of both subsistence cultivators and their uncontrolled free-ranging livestock. The qualification "uncontrolled" is the key to the problem. Methods of managing livestock that can maintain the stability of watershed slopes are known. These methods also raise the standards of living of the farmers, but they require major departures from the outdated traditions that have already brought mountain communities in India and Nepal to a state of increasingly severe poverty.

Widespread misuse of forestland is reported from the hill districts of the Indian Himalayas in the states of Himachal Pradesh and Uttar Pradesh. Of the slopes too steep for cultivation, which are officially designated as forestland, some 40 percent are in a degraded state. Overgrazed by unproductive, undersized scrub animals kept mainly to produce manure, the slopes have lost their useful tree species and carry unpalatable shrubs such as lantana, berberis, and rhus. Because most of the manure is lost on the hillside, five or six cattle are considered necessary to provide the manure for growing one-half hectare of crops.

A case study from Nepal

A well-documented project in Nepal, which I studied on an assessment mission, illustrates both the problems of livestock management and their effective technical solutions. The evidence is more valuable because the outcome has been assessed economically and is therefore given in some detail.

Nepal is a country of spectacular geography, occupying a central sector of the high Himalayas, together with a narrow strip of alluvial plains, the Terai. Under a monsoon climate with 1500 to 3500 millimeters of rainfall per year, Nepal's only major natural resource is its vast potential for hydropower. India provides a market for the export of electricity. Only 110 megawatts are as yet commissioned from Nepal's 20,000 megawatts of power-generating potential. Very large investments are under study by the two governments and the World Bank, but the vital condition for funding the construction of major reservoirs is stable soil and water conditions in the watersheds. A major advantage to India would be reduction of the annual floods that increasingly cause havoc on the Gangetic Plains.

The problem of managing steep watersheds

The Himalayas are young mountains with steep slopes and active geological erosion. The soils are stabilized by the vigorous

Steep Himalayan slopes are effectively stabilized by dense forest cover until humans intervene.

growth of forests. Because the Terai was lethally malarial until recent times, Nepal's population developed as hill farmers by clearing the forests.

As the population grew, increasingly steep slopes were cleared and terraced. Where the soil is deep, stable level terraces have been constructed, but where it is shallow, outward-sloping terraces have been made, which suffer sheet erosion until they are abandoned. Above the terraces, slopes that are too steep and rocky for even the most adventurous farmers have been gazetted by the government as protected forest land. Traditional rights to cut fuelwood and fodder and to graze livestock, however, have become destructive as Nepal's population growth has accelerated (rising from 2.1 percent in 1971 to 2.8 percent in 1979).(33)

Two-thirds of Nepal's 16 million people live by subsistence agriculture in rugged hill country, which is already seriously overpopulated. They depend on their livestock to gather nutrients from the mountain slopes because on many of the terraces, soils have become so exhausted by continuous cropping that farmers no longer attempt to plant food crops unless they can apply dung. The livestock numbers have been increased by blindly competitive overstocking of a common fodder resource, so that forage species have no opportunity to grow. Of a population of 8.5 million cattle and buffalo and 4 million small stock, 80 percent graze on mountain forestlands. The Department of Livestock Development and Animal Health estimates that the animals are starved, getting only half the dry-matter intake they need for productive growth.(24) As a result of the poor growth of small stock and the prohibition on the killing of bovines, Nepal imports more than one-half of the meat supply for its cities.(25)

The collection of dung is limited because only 40 percent is dropped in the pens at night. The rest is scattered over the mountain slopes. The hill communities were reported in a 1979 World Bank review to be producing less than 8 months of subsistence-level food supplies each year, with a declining trend. Famine relief has already been needed in the Far West region.

For watershed stability, the misuse of livestock is severely damaging. An aerial study of soil erosion in Nepal, supported by ground-truth expeditions, as part of the UNDP/FAO Integrated Watershed Management Project, found few landslides or major

Excessive grazing has reduced a formerly forested slope in Nepal to steep wasteland. In the foreground is a protected area being restored.

gullies starting from the terraced fields. The severe erosion was from the overgrazed land immediately above the cultivated zone.(15) A recent government study estimates that in mountain areas, 1.8 million hectares are in urgent need of soil conservation treatment.(29)

The critical change in land use

The Department of Soil Conservation and Water Development under the Ministry of Forests launched a project to restore watershed stability to the steep, overcrowded, and badly misused basin of the Phewa Tal, a natural lake in the middle mountains. The project was funded by UNDP and carried out by an FAO team led by an Australian forester who understood livestock as well as trees.

The first move was to consult the villagers in each of ten wards of a badly eroded panchayat (local authority area). The team proposed to employ villagers to set up nurseries and to plant fodder grasses (*Pennisetum purpureum* and *Setaria sphacelata*)

Severe gully erosion starts on Nepal's steep overgrazed slopes and cuts through terraces, destroying cropland. The scale of the damage is indicated by the small village, lower left.

and fodder legumes (*Stylosanthes guianensis*) in the major gullies near the farms. When there was fodder to cut, all livestock would be either tethered or stall-fed and free-range grazing would cease. Villagers would then be employed to plant Nepalese alders (*Alnus nepalensis*), a nitrogen-fixing fuelwood tree, on the denuded upper slopes. Villagers were also given seedlings of their favorite fodder tree, *Artocarpus lakoocha.* However, the proposed changes were too great for most of the villagers and only three of the ten wards accepted.

The results of the change

In the adopting wards, an important feature of the change was that women found the task of carrying fodder grasses from the planted gullies to be much less tiring than searching the steep mountain slopes for dry cattle dung to be carried to the terraces. Penning the cattle permitted grass to grow again on the slopes between the seedling trees. Neighbors from outside the wards were invited to cut the grass, paying a small charge for each bundle. Prunings and thinnings from the fast-growing Nepalese alders provided fuel. The soil surface was stabilized and surface erosion was greatly reduced. More of the generous 1500 millimeters of annual rainfall penetrated into the soil and a spring that had been dry for some years resumed a modest flow.

A major benefit of the changes was a big increase in manure collected from the stall-fed cattle, which made possible a winter crop of wheat to follow the summer crop of hill rice. Most of the land of the neighboring wards that had not joined the scheme remained unplanted in winter for lack of manure.

An additional bonus, arranged by the project team, was an exchange of the small hill cows, which gave only a half liter of milk per day, for buffalo cows giving 4.5 liters. The buffalo are not agile enough to graze on the mountains, but they are excellent producers when stall-fed. Because buffalo are not sacred animals, their calves could be raised and sold for meat. Goats, which are the source of ready cash to the villagers, thrived so well when tethered and fed that they fetched a far higher price in the market than goats left to scavenge. A Nepalese economist studied the farmers' household budgets in the fourth year of the scheme, after the wage payments for planting had ceased, and

Limited supplies of manure allow only one-quarter of this Nepalese village's fields to be planted to wheat. Without manure, farmers' yields would not equal their seeding rate.

found incomes to be four times those of households in neighboring wards. All seven of the wards that had earlier refused the scheme had by then applied to join.

The input that accelerated these changes was a small fund to pay villagers to replant the barren lands. These funds went to the subsistence farmers of the community, a feature that is sadly uncommon in aid schemes.

Economic assessment

A detailed economic assessment of the Phewa Tal Project(8) estimated a 20-year benefit-to-cost ratio, discounted at 10 percent, of 1.7:1. This study assumed that the change would be made gradually, with all livestock tied up in the seventh year. In the meantime, the grass-cutting areas would be protected by live hedges.

Fodder from the hills

Fast-growing fuelwood trees, close planted as in normal forestry practice, will suppress the grasses between them within 3

Buffalo are unsuited for grazing steep mountain slopes, but when stall-fed, they give nine times the milk yield of the small hill cattle.

or 4 years. One alternative is to close slopes to grazing and reserve them for cutting of grass. Broadcasting of seed of vigorous forage legumes such as "stylo" can usefully improve the forage yield and quality. Another alternative is to plant the trees at a wide enough spacing to allow grass to grow between them (see Chapter 6). This type of planting practice is outside the experience of most conventionally trained foresters and therefore needs to be part of in-service instruction for field staff.

Fodder trees are much prized by Nepalese hill farmers, but they lop them so closely that no seed is produced. In order to be able to issue seedlings of Badahar (*Artocarpus lakoocha*), the project staff had to pay farmers to protect their trees and to provide seed for the nurseries. Panday(17) lists 134 species in use by Nepalese farmers as fodder trees, with their approximate annual yields. The yields range from 50 to 200 kilograms of fresh leaves with a dry-matter content of 25 to 35 percent. For feeding livestock, fresh green tree foliage is an important complement to crop residues such as straw and stover of rice, maize, and millet in the dry season. The leaves provide protein and minerals, but grass or crop residues are necessary to provide carbohydrates for energy supply.

Estimates of annual yields of dry matter from the present overgrazed common lands in the Middle Mountains are less than 1 t/ha. When land is protected from grazing, some 3 t/ha of grasses and leaves can be expected in this high-rainfall regime. Areas with fodder trees spaced at 15 × 15 meters have been reported to give a sustained annual yield of about 10 t/ha of dry matter from foliage, together with about 1.5 t/ha from grasses.(13) The Lumle Agricultural Centre, a valuable experimental station and extension base developed from the British Gurkha resettlement scheme, has demonstrated the advantages of planting fodder legumes to use up residual moisture after cereal crops and the advantages of growing fodder on the terrace slopes or risers.

Foresters would prefer to see more of the terraced land used to grow fodder. This is unlikely to be a solution while the middle mountains remain short of food. Grain purchased from the Terai is carried up by porters to make up the difference. The more economic solution of transporting fertilizers has begun on a small scale.

Water for hill livestock

In the Phewa Tal project, an important development cost, only 30 percent of which economists allocate to the agricultural costs,(8) is the improvement of water supply. Some of the area has good hillside springs; the stall-fed livestock are driven to water along stone-walled tracks that protect the crops. However, where springs are far apart, so that the animals have to undertake daily descents and climbs on rough steep tracks, the milk yields of buffalo are reduced.

Although hand carrying of fodder to stall-fed livestock is an established practice, the hand carrying of water in steep hill country is not an acceptable labor. Fortunately, modern technology is rapidly providing solutions. Small stock ponds can be waterproofed by buried plastic sheeting, but they must be fenced and provided with drinking troughs to prevent damage from trampling by livestock. Thick-walled plastic piping offers a relatively cheap and technically simple system of conveying water over the short distances involved in well-watered mountains. For village water supplies, larger sources are necessary. In the Phewa Tal project, a 17-kilometer pipe was laid. Together with several effective designs of hand pumps from the WB/UNDP Global Hand Pump Project in the U.N. Drinking Water Decade, this new technology is encouraging. Improvement of water supplies for livestock on steep land, however, should be undertaken only if stall-feeding is accepted. Otherwise it may lead to even more intensive overgrazing.

The scale of changes needed

Technical solutions are thus available to change the effect of livestock from accelerating environmental destruction to contributing to higher standards of living for hill people. Surging population growth in the tropics is removing all other options. The planting of fuelwood and fodder has been recognized by the World Bank and the U.N. as critical for the future of the Himalayan countries. A report by the Asian Development Bank in 1982(2) estimated that Nepal needs to plant fuelwood at a rapidly increasing rate that rises to 50,000 hectares per year by 1990. This is in contrast to the 11,600 hectares planted during the previous 10 years. Nepal's Sixth Plan 1980/81–1985/86 included planting of 84,000 hectares of trees. A World Bank report on options in the energy sector(32) calls for a massive reforestation program to achieve regular output from 1.2 million hectares of fuelwood plantations by the year 2000. The interlinked crises in supply of fuelwood and fodder can only be met by mobilizing the hill communities.

The legal basis has already been established. The Ministry of Forests and Soil Conservation can transfer to each village community the land rights for 500 hectares of destroyed forest and 125 hectares of surviving forest, together with the responsibility for directly managing these areas. The areas are being agreed upon, with management programs provided by the district forest officers. It is important that the livestock fodder aspect should be fully recognized by the foresters in this urgent campaign.

The Asian Development Bank report draws attention to the basic issue of the balance of supply and demand in Nepal.

Although the supply side can be improved by better land use for crops, livestock, and fuelwood, the demand side needs urgent attention from family planning services. "A simultaneous and complementary program aimed at reducing the rapid population growth rate, which has created intense pressure on the resource base of the country, is absolutely essential."(2)

Livestock on watersheds under subtropical pine forests

The same problems of misuse of steep land on important watersheds are seen at high altitudes, where temperatures do not permit fast-growing tropical fuelwood or fodder. An urgent case is that of Pakistan's only two major reservoirs, the Tarbella Dam on the Indus River and the Mangla Dam on the Jhelum. The watersheds of both reservoirs in the mountainous Hazara division of the North West Frontier province contain critically overpopulated areas. A watershed management and reforestation pilot scheme was begun in the 1960s and in 1977 was extended to the whole 1,274,000 hectares of the Hazara division, with major support from successive Food-for-Work projects of the World Food Program. Nonetheless, the sedimentation rates of both reservoirs continue to increase and their expected storage life has been seriously reduced (although the volume data appear subject to debate about the consolidation of sediments).

The economy of Pakistan depends entirely on the irrigation of the alluvial plain of the Indus River, but the total storage provision is very small indeed. The Tarbella and Mangla dams together provide only about 5 days supply. Another dam on the Indus, below Tarbella at Kalabagh, is under discussion. The Indus rises some 1500 kilometers away in Tibet and drains a total basin area of 1,165,000 square kilometers. Most of this area lies outside Pakistan's borders. The sediment transport from the upper watershed is believed to be mainly of glacial origin, whereas the heavy input of soil eroded from the lower watershed comes mainly from within Pakistan's borders. The most seriously damaged area is in the steep Hazara pine forests in which the Food-for-Work projects concentrated on 200,000 hectares.(11)

Detailed legislation has enabled the Forest Department to

regulate commercial felling of the forests of Chir pine (*Pinus roxburghii*), but the complex tribal pattern of landholdings has made control of the farmers and their livestock far more difficult. The Food-for-Work projects persuaded owners to close 28,000 hectares to grazing and to reforest them with Chir pine. Because the trees grow very slowly, the owners can harvest the grass between them for about 15 years. The cut grass is fed to livestock as a supplement to forage from crop residues of straw and stover.

Clearly the replanting of about 14 percent of the area cannot have much direct effect on the sedimentation of the reservoirs, but a campaign of soil conservation and crop improvement has probably had a greater effect. However, fuelwood collection by lopping and felling in the natural forests still takes a heavy toll. It seems probable that Pakistan's major watershed problem will continue to increase until it attracts a national effort commensurate with its importance to the national irrigation system.

A sad complication has resulted from the war in neighboring Afghanistan. Across Pakistan's mountainous frontiers, tribal people fleeing from aerial bombardment brought some thousands of livestock, which remained in the high forest during the monsoon. International relief organizations have begun some reforestation schemes to remedy the damage.

Livestock above the tree line

Above the Himalayan tree line at about 4000 meters, rock and ice dominate the watersheds, but there are substantial areas of alpine grasslands. The rainfall is low (300 to 500 mm), and the growing season lasts only 4 to 5 months. In Nepal, Sherpa tribes move up into the pastures in spring and retreat to the live-oak groves (*Quercus incana* and *Q. lanata*) of the upper forest in winter. The oak foliage is the main winter forage. Some of these areas are already overstocked and overgrazed.(1, 4) Mustang in Nepal's Western province and Dolakha in Central province still suffer from the influx of refugees and their livestock from Tibet in the 1960s, although some areas of this vast mountain territory are as yet underused.(25)

In the alpine areas of Dolakha, severe ecological changes have resulted from overcutting and overgrazing, mainly in the last 30 years. Alirode(1) ascribes the damage to increasing numbers of livestock driven up into the high pastures from the areas of overgrazed forest below. Few trees are now found above 3500 meters; the live-oak forests have been cut to extinction, but the population density has not been high enough to prevent some natural repair of the watershed surface. The overgrazed grasslands have been invaded by the thorny and unpalatable shrub *Berberis insignis,* and the live oaks have been replaced by dense thickets of the inedible shrub *Rhododendron arborem*.

Despite the damage, there are no hydrological hazards for the downstream community because the rainfall is too low to generate serious floods. The purchase of excess livestock and some employment provided by the tourist industry seem to be the only forms of help that can reach these remote areas.

Livestock on lowland watersheds

On lands of low relief, well-managed grazing can provide stable and profitable patterns of land use for tropical watersheds. Under uncontrolled communal grazing, however, the land rapidly sheds water from heavy tropical storms. The overland runoff is less damaging than it is on the hills, but massive sediment transport can still occur (see Chapter 2).

In humid tropical climates, the clearing of heavily forested lands, even in mild topography, rarely produces economically viable livestock enterprises. The flat, wet, heavily forested basin of the Amazon provides some sorry examples on a large scale. The difficulties are those long suffered by shifting cultivators in the forests. Under the intensive leaching conditions of heavy rainfall and high temperatures, most of the available plant nutrients are found in the vegetation rather than in the soil. The nutrients are returned to the soil as leaf litter, which decays rapidly, releasing them to the leaching of warm rainfall. The nutrients are promptly retrieved by the dense surface mat of roots, thus circulating in an effectively closed cycle. There are heavy losses from this limited supply of nutrients when the forest is cleared and burned. Pasture maintenance is difficult, and productivity falls rapidly during 4 or 5 years.

Replacement of tall forest by grassland also causes a substantial reduction in water use because both interception and transpiration are reduced. Slow drainage in the flat terrain results in swampy conditions. Although profits may be made from livestock enterprises in the first few years while the residual soil fertility is used up, these ventures have demonstrated on a large scale that wet forested land is unsuitable for ranching. In Brazil's wetter areas, more than half a million hectares of pastures cleared from forest are degraded and overrun by unpalatable weeds.(28) Where there is a strongly contrasting dry season, however, the climax vegetation is open forest, from which Brazil has successfully developed 2 million hectares of pastureland and a major beef industry.

Livestock on drylands

The dry tropical savannas of East and Central Africa, tropical Australia, and South America provide many examples of successful watershed management by grazing of livestock. But they also contain vast and growing areas that have been seriously degraded by overgrazing. In the countries of the African Sahel, watershed degradation has already resulted in famine.

In contrast to the high rainfall hill country already discussed, where the technologies are known and the problems are essentially those of breaking away from untenable traditions, the difficulties of livestock management in dry tropical lowlands are less easily solved. These areas suffer the dry extremes of the erratic patterns of the ITCZ in which rainfall for 2 or 3 years in succession may be inadequate to grow food crops or to maintain pastures. This is well illustrated by the century-long rainfall record of the Sahel, which is shown in Figure 25 as a percentage of the long-term average.(9) The irregular groups of wet years and dry years show no periodic cycles and no long-term trends.(12) Recent years have raised questions about a possible change of climate, but geological records indicate that, after changing repeatedly 2 to 3 million years ago in the Late Quarternary, the climate of the Sahel has been relatively constant for the last 2500 years.(14)

Desertification

Local climatic effects are, however, well established (as already noted in the discussion of windbreaks in the previous

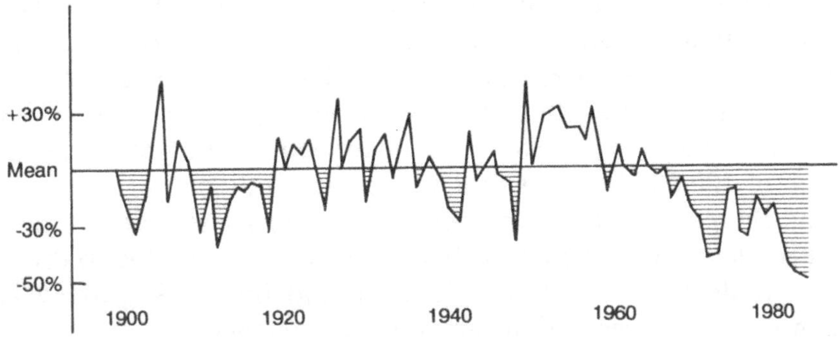

Figure 25. Rainfall in the Sahel: departures from the long-term average. Statistically random rainfall sequences may show large cumulative departures. *Sources:* National Academy of Sciences, *Environmental change in the West African Sahel* (Washington, D.C.: National Academy Press, 1984); C.K. Folland, D.E. Parker, and T. N. Palmer, 1985, Sahel drought and worldwide sea surface temperatures (1910–1985), *Nature* 320:602–607.

chapter). Removal of trees for firewood increases wind speed over the ground surface. Overgrazing strips away the vegetation, and livestock trample the soil to a dusty condition. Dust storms and wind erosion follow, and the exposed soil surfaces reach high temperatures. Hot, dry winds, laden with abrasive dust, then desiccate the vegetation of areas downwind.

The proportion of incoming radiation reflected by the land surface is termed the *albedo.* A tall forest cover absorbs most heat and reflects only about 10 percent; a complete grass cover reflects about 25 percent; and a bare soil reflects up to 40 percent. On a global scale, the rise in albedo of the overgrazed areas of the tropical and subtropical drylands is already serious enough to have a possible effect on world climate. The increasing reflection of solar radiation reduces the heating of the land surface. This effect is, however, obscured by the opposite effect of the carbon dioxide added to the atmosphere by increased burning of both fuelwood and the fossil fuels (coal, oil, and gas). The extra carbon dioxide, although transparent to incoming shortwave radiation, prevents the escape of longwave back radiation, as does the glass in a greenhouse. The evidence is, as yet, inconclusive. Opposing views, based on different computer models, are expounded by Sagan et al.(27) and by Potter et al.(23)

That overgrazing strongly affects reflection back to outer

space is well illustrated by satellite imagery, which clearly shows the boundary of protection by the border wire of the Negev against Bedouin grazing practice beyond; similar contrasts can be seen in the pictures of the boundaries between South Africa and overgrazed Lesotho.(16) A green pentagon seen on a NASA satellite image of the Sahel was identified on the ground as an area of 100,000 hectares that was being rotationally grazed behind the protection of a barbed-wire fence. These pictures indicated clearly that the desertification process is not due to drought. If destruction by woodcutting and livestock grazing is controlled, the vegetation survives the droughts.

The paradox for range management in the tropical drylands is that during several years of good rainfall a precarious imbalance is built up. More calves survive, cattle numbers increase, and sheep and goats proliferate. Without warning, a season of poor rainfall can halve the carrying capacity of large regions, and the dry phase may persist, as Figure 25 demonstrates. The term *drought* is used in wetter climates to describe an unusual event. In the climate of the Sahel, it is an inescapable feature. The true carrying capacity of the drylands is therefore that of their dry years. For land-use planning, averages are dangerous; the worst-case scenario determines survival.

When livestock populations were small, herds survived dry spells by ranging wider to graze a thinner crop of grasses from a greater area. With present livestock densities in the Sahel, there are no unused areas to act as reserves. In Australia, such problems are familiar and are now met by the use of highly specialized transport—heavy trucks and trailer trains—to remove stock to better pastures or to large abattoirs. Nonetheless, emergency slaughter and burial of starving sheep have still been needed in recent years. Australian evidence presented to the U.N. Conference on Desertification(31) made the point that in northwestern Australia, government intervention occurred in drought years, whereas the recovery of overgrazed drylands occurred in the good years. In the sheep-raising area, which has an erratic 200 millimeters of annual rainfall, the conduct of livestock management in the good years determines whether the rangeland continues to be productive or degrades to desert. Government help in the good years would be more productive and less expensive than disaster relief in drought years.

Unless the grassland is relieved of the starving livestock, the soil surface is stripped and trampled and the palatable herbage species are destroyed or so weakened that they are invaded by unpalatable weeds. Perennial grasses are replaced by short-lived annuals, while unpalatable woody species such as *Lantana camera* increase. Violent tropical rainstorms are increasingly shed as overland runoff from the trampled surface, even on slopes of only 2 or 3 percent, and sheet erosion carries off the topsoil. Flash floods then replace infiltration onto the soil; groundwater aquifers are not recharged, so that springs dry up and streamflows fail in the dry season. This is the technical process of desertification, which refers not only to the spread of existing deserts but also to the degradation of ecologies to desert conditions.

Subsistence graziers add to their own problems by the use of fire to remove the bulk of dry herbage in order to induce new growth. Fires are also started by hunters to drive game out of cover. Bush fires have been a feature of the African continent throughout recorded history; documents by Carthaginian sea-traders reported fires visible to coasting ships in the 5th century B.C. Such fires volatilize organic nitrogen compounds; nutrients in the ash from vegetation and burned dung are lost by leaching. The actual loss of forage, which although of low digestibility will keep ruminants alive, amounts to a significant reduction in carrying capacity.

Far from diminishing the effects of desertification in Africa, many Western aid projects have added to the problem through their short-term outlooks. In the Sahel, the drilling of large numbers of boreholes as emergency aid has prolonged the maintenance of livestock in numbers far greater than the carrying capacity of the land.

In 1977, the U.N. Desertification Conference estimated that 57 million people were in serious difficulties as the result of the destruction of their rangelands by overgrazing. By 1984, the numbers had reached 135 million, according to a somber report by the U.N. Environmental Programme (UNEP) on the continued increase in the desertification of the world's tropical and subtropical drylands.(30) UNEP estimated that, worldwide, 850 million people are at risk from deterioration of drylands. The combination of population increase and the misuse of resources is most acute in Africa south of the Sahara, but it is also destroying the future of millions in Andean South America, Mexico, and parts of southern Asia. Land degraded to desertlike conditions, some of it irretrievably lost, is increasing by 6 million hectares annually. Land degraded to zero productivity, much of which could be restored by better management, is estimated to be increasing at 21 million hectares per year. International efforts to arrest this catastrophic degradation of resources will be needed ever more urgently as populations increase. So far these efforts have been inadequate and ineffective. UNEP observed that "field-oriented projects to arrest desertification processes are in a minority, with a tendency for the funds directed to desertification control to dwindle as they move downwards through the administrative machinery."(30)

The tsetse-fly problem

A potentially great addition to the total area of desertification in Africa is implicit in the increase in aid projects devoted to control of the tsetse fly. At present 10 million square kilometers of tropical Africa, or about one-half of the inhabited area of the continent, is either lightly stocked or denied altogether to cattle by the disease trypanosomiasis (Fig. 26). The trypanosomes are carried by tsetse flies, a group of vector species of the genus *Glossina.*

Very large programs of bush clearing to remove the fly's breeding habitat, together with spraying of insecticides, are steadily enlarging the grazing areas. Unfortunately, the history of tsetse-fly control has been one of overgrazing and erosion of the cleared land. The more recent project agreements usually contain a formal undertaking by the recipient countries to control grazing in the cleared areas, but this has proved to be a nominal gesture. From the viewpoint of the watershed manager, misuse of the cleared land is usually corrected by regrowth of trees. Pioneer species form dense thickets unless a grazing rotation is organized to provide grass for a controlled fire at intervals of 3 to 5 years. The reestablishment of thicket growth restores the tsetse habitat. This has occurred on a major project area in Uganda.(10)

Many scientists and administrators concerned with land management problems in Africa have expressed fears about the potential outcome of a massive program against trypanosomiasis

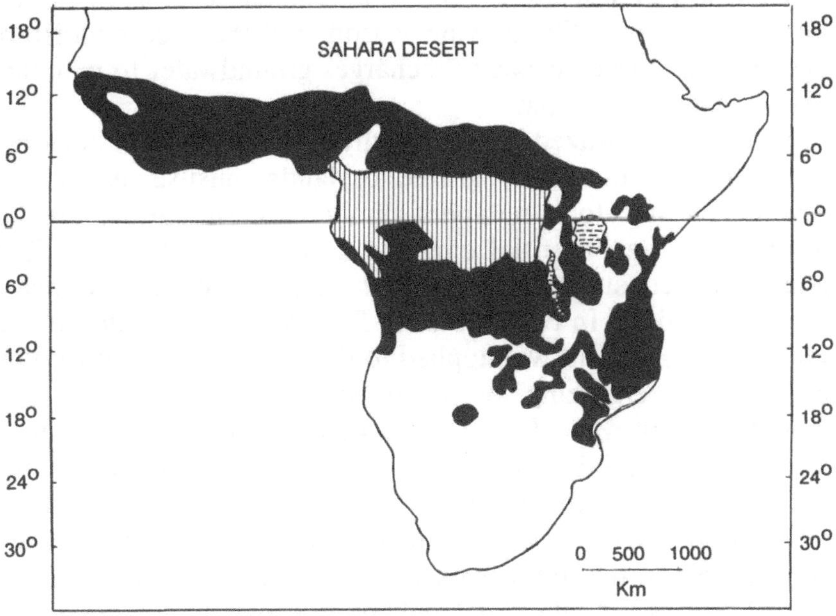

Figure 26. Areas infested by tsetse flies. The habitat of the tsetse fly, vector of trypanosomiasis, a fatal disease of cattle, is the woodland or savanna below 1600 meters elevation (areas in black) and the rain forest (shaded).

funded by the European Community. The campaign will attack tsetse fly in four African countries by spraying the toxic chemical endosulphan from the air. The result could be a sharp increase within only 10 to 15 years in the number of people in distress from land misuse in these countries and a disastrous deterioration of their watershed stability. Clearly, the tsetse fly does need to be eliminated, but a substantial part of the investment in any such project should be devoted to organization and control of the grazing potential thus released. This step is, however, a low priority for the governments concerned and is likely to prove even more difficult than the control of trypanosomiasis.

Restoration of overgrazed rangeland

As with other types of land use, the proper management of rangeland watersheds is required for productive farming. Good land management benefits the farmers as well as the downstream communities. Maintenance of a vigorous grass cover prevents soil erosion and secures infiltration. Increased storage of soil moisture prolongs the growing season, and the excess moisture beyond soil storage capacity recharges groundwater to maintain dry-weather streamflow.

Severely overgrazed land is usually the result of overstocking; however, in communally grazed drylands, misuse may be so severe that livestock numbers, limited by starvation, are far lower than the potential carrying capacity of the land. The grasses are consumed before they have grown enough to give a productive yield. In two studies in East Africa in which rational grazing management was applied to the wastelands produced by communal overgrazing, both the numbers of cattle and their condition improved. That is, the number of animals fattened exceeded the number that had been starving on the same area previously. Since quantitative information from the overgrazed drylands of Africa is difficult to find, the two experiments are summarized in the following sections. The first was in a densely settled area of the hills of Ukambani in Kenya where food crops and livestock were equally important. The second was in the upper watershed of the Nile, in the Karamoja district of northern Uganda. Here livestock dominated the rather spartan economy.

Kenya: The Makaveti experiment

In the area of the Kenya experiment,(20) Wakamba subsistence cultivators grazed cattle, sheep, and goats on communal areas of eroded thornscrub. The site, Makaveti, was at an altitude of 1500 meters. The annual rainfall received during the 7 years of the experiment ranged from 250 to 1350 millimeters, with a mean of 660 millimeters distributed in two seasons—the "long rains" and the "short rains." The soils were red, basement-complex sandy loams.

The thornscrub was hand cleared and fenced into 80 quarter-hectare paddocks set out on the contour. The five pasture treatments were (1) resting 2 years to allow recovery of the natural veld grasses, for comparison with contour plowing and sowing with (2) African foxtail (*Cenchrus ciliaris*) or (3) Rhodes grass (*Chloris gayana*), or planting with (4) star grass (*Cynodon dactylon*), or (5) Makarikari grass (*Panicum coloratum* var. *makarikariensis*). The design was four replications of a 5×2

Above, communal grazing land in Kenya before clearing and restoration of grass cover. Below, established pastures planted to Makarikari grass.

factorial with split-plot fertilizer application. Half of the paddocks received 325 kg/ha of NPK fertilizer. Half of the main plots were closed late in the long-rains season for haymaking.

Grazed by groups of eight small-zebu bullocks averaging 230 kilograms liveweight, the treatments gave carrying capacities during 7 years ranging from 0.6 hectares to 1.5 hectares per animal. Plowing proved to be unnecessary because the natural veld grasses recovered within 2 years to give equally good yields. The volunteering grasses included *Cenchrus ciliaris, Panicum maximum,* and *Themeda triandra.*

Haymaking proved to increase yields but was unreliable during the 5 years in which it was tested. Yields of 2.5 t/ha in 2 good years, 0.7 t/ha in 1 year, and no hay at all for 2 years illustrated the vulnerability to rainfall fluctuations. The hay maintained the cattle in excellent condition in the dry season in spite of its coarse quality. The application of fertilizer raised hay yields by 60 percent, but it increased grazing days significantly on only the star grass treatment. Fencing was needed to give precision for experimental purposes, but because livestock in this area were normally grazed in small groups, all of which were herded all day, fencing was not essential to pasture management.(20)

In subsequent test cropping with maize, sorghum, and interplanted beans, yields were doubled by the residual effects of the fertilizer, but there was little difference among residual effects from the pasture types. Crops were poor on the unfertilized areas; even under rational grazing management, buildup of fertility in these eroded soils was very slow unless some of the lost nutrients were replaced.(21)

This experiment gives encouragement to the watershed manager because it demonstrated that stability could be restored and yields raised by methods that lay readily within the means of the Wakamba farmers. The low productivity and the destructive soil erosion were due to lack of organization and management rather than to poverty of resources. (It is sad to report that when this area was revisited 20 years later in 1980, the population had increased substantially and appeared to be in excess of the area's carrying capacity for both people and livestock. This is part of Kenya's intensifying problems from a population that doubles every 17 years.)

In experiments in Kenya, hardy small Zebu cattle thrived on coarse hay, but erratic rainfall patterns made haymaking unreliable.

Uganda: The Atumatak experiment

The Uganda experiment was conducted in an area of open savanna woodland that had a single annual rainy season and a 7-month-long dry season. The average annual rainfall was 500 to 750 millimeters. The able-bodied men of the Karamajong followed nomadic seasonal patterns, returning to a fixed base in the wet season. In the dry season, the livestock were trekked to hot lowlands, while the women, children, and the elderly remained with the milking cows and goats. Although this area could have been managed as productive rangeland, many hundreds of square kilometers were reduced ecologically by overgrazing to thornscrub, with annual desert grasses replacing perennials. Overgrazing was so severe that flash floods repeatedly damaged bridges on the sparse road system. An administrative policy to improve water supplies by drilling boreholes and building small earthen dams was formally suspended until grazing control could be improved because each new watering point initiated a new zone of devastation.

An intensive study of the watershed regime and of the hydro-

A typical area of overgrazed savanna in East Africa has been reduced to thornscrub and annual grasses.

logical effects of grazing management was therefore undertaken by the East African Agricultural and Forestry Research Organization in cooperation with the water engineers, agricultural officers, and administrative officers of the Uganda government. A flat shallow basin of 8 square kilometers, divided by a ridge of rock into two approximately equal watersheds, provided two stream channels that delivered spate flows for 2 hours after the characteristically heavy rainstorms. To establish the experiment, much consultation was necessary to overcome the suspicions of the herders; a ritual present of two heifers to the headman and patient explanation of the reasons for fencing and for counting of stock eventually secured a measure of cooperation.

Quadrat sampling recorded 40 percent of the soil surface to be bare for 9 months in each year. Ironically, the place name Atumatak means an area for fattening calves. The two watersheds were fenced with brushwood, and grazing continued. A daily count of cattle, sheep, goats, and donkeys for a year gave a stocking rate, in adult-cattle equivalents, of 1 beast per 4 hectares.

Dense grass that volunteered when the watershed was cleared of scrub and closed to grazing for one year. A survey found 37 perennial species of grasses and 22 annual species.

Heavy concrete flumes were constructed to measure flows up to 30 m^3/sec, and calibrations were checked with current meters. Sediment deposition gave frequent difficulties. A rain gauge network and meteorological station were established, and penetration of rainfall into the soil, to a depth of 2 meters, was recorded electrically at 20 sites.

After 4 years of observations, one watershed was cleared of bush and closed to grazing for a year. Grazing continued on the control catchment. Grasses recovered with unexpected rapidity on the cleared area. Perennial species had survived in thickets of thornbush and no reseeding was needed. But on the steeper slopes where topsoil had been lost and gravel lay on hard subsoil, recovery was still observed to be slow after 7 years. A botanical survey on 30 transects of 0.4 hectares recorded 59 species of grasses, of which 37 were perennials.

Managed grazing resumed 1 year after closing, with mature cattle at 1 beast per 4 hectares. The cattle were fattened, sold, and replaced; the carrying capacity increased to an estimated 1 beast per 2.4 hectares (6 acres).

Records of soil moisture availability, analyzed throughout 11 years, showed that rainfall consistently penetrated deeper under the managed grazing: Soil was wetted to 1-meter depth twice as often as soil under the tribal pattern of grazing. The grass cover assisted infiltration and increased the water use by transpiration, so that surface flows decreased. Peak rates of stormflow from the treated watershed were reduced to about one-half of those from the control.

The validity of these results was then demonstrated on a large scale as Africa's turbulent history overtook the scientific work. The experiment ended in 1975 when raids by neighboring tribes killed many of the Karamajong herders. The rest retreated from the area. Grass cover then increased to fuel dry-season fires, which thinned the thornbush so that the tall perennial *Hyparrhenia* grassland replaced much of the thicket. Some hundreds of square kilometers began a slow return to a savanna ecology.

Lessons for watershed management are that overgrazed drylands can be restored to soil stability by correction of grazing management; reductions both in stormflow and in total water yield can be expected. Full details of soils, climate, vegetation, and hydrological data are in publications by Pereira et al.,(22) Pereira,(19) and Blackie, Edwards, and Clarke.(3)

Traditional methods of grazing control

In Mediterranean drylands, an encouraging return to traditional patterns of grazing management has been reported despite rapid increases in populations of both graziers and their livestock. According to Draz,(5, 6, 7) the ancient Arabian system of *hema* has been revived in the Syrian steppe. Under the system, a committee of tribal elders declares each of a series of low hill ranges to be successively closed to grazing, usually for 5 years. If a shepherd takes his flock into the closed territory his penalty is a fine of two sheep, which he has to kill and cook and serve to the committee in his own tent. Draz reports a similar system from the remote areas of the dry rangeland of Saudi Arabia, where the closed area is available for hunting, with any straying livestock legitimately included in the bag.

In the Mediterranean drylands, an urgent extension problem is to interest cattle-owning communities in the more profitable methods of managing their rangelands. Practical adaptation of the *hema* system to a modern cooperative organization with fattening yards and marketing facilities is receiving government and international support in Syria.

References

1. Alirode, Philipe. 1983. *Transhuming animal husbandry systems in the Kalingchowk region.* Kathmandu: Swiss Association for Technical Assistance.
2. Asian Development Bank. 1982. *Nepal: Agricultural sector strategy study.* 2 vol. Bangkok.
3. Blackie, J.R.; Edwards, K.A.; and Clarke, R.T. 1979. Hydrological research in East Africa. *East African Agriculture and Forestry Journal.* 43 (Special Issue). 313 p.
4. Brown, R.W. 1982. *Recommendations for revegetation and management of denuded lands in Mustang.* Kathmandu: U.S. Agency for International Development, Southeast Consortium for International Development.
5. Draz, O. 1969. *The "hema" system of range reserves in the Arabian peninsula.* FAO/PL 13. Rome: FAO.
6. Draz, O. 1975. *Report on mission to countries of the Arabian peninsula.* Rome: FAO.
7. Draz, O. 1977. *Role of range management in the campaign against desertification: The Syrian experience.* UNCOD/MISC13. Damascus: UNDP Regional Office.
8. Fleming, W.M. 1983. Phewa Tal catchment management program: Benefits and costs of forestry and soil conservation in Nepal. In *Forest and watershed development and conservation in Asia and the Pacific,* ed. L.S. Hamilton. Boulder, Colo.: Westview Press.
9. Folland, C.K.; Parker, D.E.; and Palmer, T.N. 1985. Sahel drought and worldwide sea surface temperatures (1910–1985). *Nature* 320:602–607.
10. Ford, J. 1971. *The role of trypanosomiasis in African ecology.* Oxford: Clarendon Press.
11. Khattak, G.M. 1983. The watershed management program in Mansehra, Pakistan. In *Forest and watershed development and conservation in Asia and the Pacific,* ed. L.S. Hamilton. Boulder, Colo.: Westview Press.
12. Mason, B.J. 1977. Man's influence on weather and climate. *Journal of the Royal Society of Arts* 125:150–165.

13. Mauch, S., and Schwank, O. 1979. *Estimating the impact of IHDP programme components.* Zurich: Swiss Association for Technical Assistance.
14. National Academy of Sciences. 1984. *Agroforestry in the West African Sahel.* Washington D.C.: National Academy Press,
15. Nelson, D.O.; Laban, P.; Strestha, B.D.; and Kandel, G.P. 1980. *A reconnaissance survey of major ecological land units in Nepal and their watershed condition.* NEP/74/020, Rome: FAO.
16. Otterman, J. 1974. Baring high-albedo soils by overgrazing: An hypothesized desertification mechanism. *Science* 186:431–433.
17. Panday, K.K. 1982. *Fodder trees and tree fodder in Nepal.* Birmensdorf: Swiss Federal Institute of Forestry Research.
18. Pereira, H.C. 1968. *Soil erosion in Ethiopia and proposals for remedial action.* Addis Ababa: Institute of Agriculture Research.
19. Pereira, H.C. 1973. *Land use and water resources.* Cambridge: Cambridge University Press.
20. Pereira, H.C., and Beckley, V.R.S. 1952. Grass establishment on eroded soil in a semi-arid African reserve. *Empire Journal of Experimental Agriculture* 21:1–15.
21. Pereira, H.C.; Hosegood, P.H.; and Thomas, D.B. 1961. The productivity of tropical semi-arid thornscrub country under intensive management. *Empire Journal of Experimental Agriculture* 29:269–286.
22. Pereira, H.C.; McCulloch, J.S.G.; Dagg, M.; Kerfoot, O.; Hosegood, P.H.; and Pratt, M.A.C. 1962. Hydrological effects of changes in land use in some E. African catchment areas. *East African Agriculture and Forestry Journal.* 27 (Special Issue). 131 p.
23. Potter, G.L.; Ellsaesser, H.W.; McCracken, M.C.; and Ellis, J.S. 1981. Albedo change by man: Test of climatic effects. *Nature* 291:47–49.
24. Rajbhandary, H.B., and Pradham, S.M.S. 1981. Livestock systems in hill farming. In *Seminar on appropriate technology for hill farming systems.* Kathmandu: HMG Ministry of Food and Agriculture.
25. Rajbhandary, H.B., and Shah, S.G. 1981. Trends and projections of livestock production in the hills. In *Nepal's experience in hill agricultural development,* 43–58. Kathmandu: HMG Ministry of Food and Agriculture.
26. Randhawa, N.S. 1980. Watershed development in India: An overview. In *Proceedings of the national symposium on soil conservation and water management in the 1980's.* Dehra Dun: Indian Association of Soil and Water Conservation.
27. Sagan, C.; Toon, O.B.; and Pollack, J.B. 1979. Anthropogenic albedo changes and the earth's climate. *Science* 206:1363–1368.

28. Serrao, E.A.S.; Falesi, I.C.; de Veiga, J.B.; and Teixeira Netto, J.F. 1979. Productivity of cultivated pastures on low fertility soils in the Amazon of Brazil. In *Pasture production in acid soils of the Tropics.* Cali, Colombia: Centro Internacional de Agricultura Tropical.
29. Strestha, B.D.; Van Ginnekan, P.; and Sthapit, K.M. 1983. *Watershed condition of the districts of Nepal.* Kathmandu: Ministry of Forests and Soil Conservation.
30. United Nations Environmental Programme. 1984. *General assessment of progress in the implementation of the plan of action to combat desertification (1978–84).* Nairobi.
31. UNESCO. 1977. *U.N. conference on desertification, Nairobi.* Paris.
32. World Bank. 1983. *Nepal: Issues and options in the energy sector.* Report 4474 NEP. Washington, D.C.
33. World Bank. 1984. *World development report 1984.* Washington D.C.

8
Watershed techniques for rainfed crops

The mechanisms of soil erosion

The physical causes and mechanisms of soil erosion have only been understood for the last 50 years out of humankind's 7,000-year-long efforts to grow food crops. The main advances have come from the U.S.A. because the rapid opening of North America to agriculture in the 19th century caused soil erosion problems on a national scale. At the same time, substantial resources were provided for agricultural science and for its practical application because farming was a source of wealth and prestige.

To reduce the damage from overland flow, the U.S. Soil Conservation Service developed systems of contour plowing with the organization of surface drainage into grassed channels on a sub-watershed basis. These techniques greatly reduced soil losses, but effective protective systems were not developed until detailed research had clarified the effects of raindrop impact on the soil surface. E.K. Edgerton's slow-motion films of raindrop impact, using gas-discharge flashes of 1/200,000 second, together with sprinkler studies of drop size and velocity by Laws(12) and studies of the protective effects of straw mulches by Duley(2), laid the foundations for conclusive studies by Ellison.(4) He measured the kinetic energy of the raindrops, the soil detachment that resulted when the raindrops struck the soil surface, and the transport of soil in runoff from a range of soil types. Ellison showed that striking raindrops act like miniature bombs, detaching soil particles and throwing them into the runoff

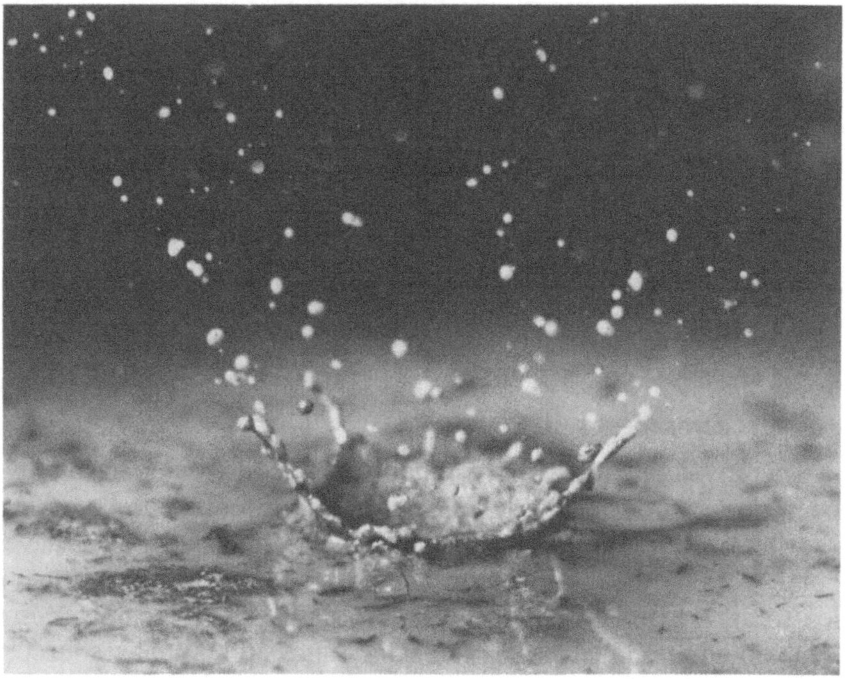

A large raindrop strikes bare soil with the effect of a miniature bomb. This is the critical point at which soil erosion begins. *Source:* U.S. Soil Conservation Service.

stream. Ekern showed that by impact on the flowing sheet of water, the raindrops cause turbulence, which adds greatly to the rate of soil transport.(3)

These experiments were done in temperate latitudes, but in the tropics a higher proportion of the rainfall characteristically occurs as high-intensity convectional storms. The erosive effects are therefore increased. That the same principles apply under tropical conditions was demonstrated by Hudson(8) in an elegantly simple experiment in Zimbabwe. Suspending mosquito gauze a few centimeters above the surface of a cultivated bare soil on a 5 percent slope 27 meters long, he showed that the soil loss was reduced to only 1 percent of that from the unprotected slope. The basic principles of soil conservation are therefore the protection of the soil surface and the safe routing of overland flow.

Protection of the soil surface

In order to protect the soil from direct assault by raindrops, the farmer must achieve a complete crop canopy. A high plant population and vigorous plant growth not only protect the soil surface, the resulting thick mat of roots reinforces the topsoil. A dense crop stand is also necessary for high yields, so that soil protection is most effectively assured by good farming standards. Leaving crop residues on the surface or mulching with grass or straw protects the soil and conserves moisture. Poor crops give sparse cover.

Minimum tillage

Minimum tillage began with the use of wide shallow blades to leave crop residues on the surface instead of turning them in with the plow. This is an effective method of soil protection, but unfortunately, in subsistence agriculture, the stalks of maize or sorghum are usually needed for forage or fuel, and they will certainly be used when livestock are stall-fed. A traditional form of tropical minimum tillage, which is used when the soil is loose enough or moist enough, is to drop seeds into holes made with a pointed planting stick. This technique is characteristic of shifting cultivation in forests. In settled cultivation, it is widely used for the planting of beans in cereal stubble.

Poor crops provide sparse protection from heavy rainfall.

In technologically advanced countries, minimum tillage is done by replacing cultivation with application of herbicides to kill existing vegetation. Then seed is sown in slots cut into the ground with heavy disk coulters. In tropical and subtropical countries that have a high standard of technological management, this is an effective method for controlling erosion. It is successful in Australia and in parts of southern Africa, but it is not suitable for small-scale subsistence agriculture. One serious hazard is the lack of safe storage space to keep toxic chemicals out of the reach of children in simple family homes; other drawbacks are the need for protective clothing for field-workers and the need for heavy seed drills.

Soil structure

Preparation of a seedbed for most cereals requires a bare soil with a loose tilth exposed to the heavy thunderstorms with which the tropical rainy seasons usually begin. It is therefore essential to maintain the soil in a receptive structural condition in which it will resist the sealing of its surface. In soil of good structure, the mineral particles are aggregated into soil crumbs held together by organic matter. This is mainly in the form of humus, the decomposed residues of plant tissues. At the high temperatures reached by soil surfaces in the tropics, the organic matter is rapidly lost by oxidation, but the vigorous root systems of tropical grasses, including cereal crops, leave large volumes of root fragments in the tilled soil. Undecomposed root fiber is of great importance in maintaining soil structural stability under tropical conditions. It even plays a positive part in the base exchange surface for plant nutrients.(7) This is another strong reason for closely spacing crop plants and for maintaining a continuous grass cover on fallow land. When studied under tropical conditions, soil reinforcement by root fragments was found to give good protection against soil erosion in the first season after plowing grassland, to give less protection in the second year, and to have negligible effect in the third year of tillage.(14)

Manuring

The most direct action that the subsistence farmer can take to maintain soil structure during the weeks before the crop provides cover is to replace lost organic matter with farmyard manure or compost, which is also needed to provide plant nutrients (as discussed in Chapter 7). When manure is being accumulated prior to application, it should be covered with, for example, a rough thatching to prevent leaching by rainfall. When being applied in the field, manure should be dug in and immediately covered with soil. Otherwise there will be serious losses of precious nitrogen, escaping as ammonia gas.

Control and routing of runoff

However good the soil structure may be, intense tropical storms will deliver rainfall faster than bare soil can accept it, so that some runoff over the surface will occur. If the structure is poor and dusty from overcultivation or trampling, the surface may be sealed by dispersed soil particles so that most of the rain will be shed as runoff. Clearly the best precaution is to have the soil surface nearly flat, with a slight gradient to guide the runoff into channels for safe disposal. However, the future livelihoods of hundreds of millions of hill-dwelling subsistence cultivators who have no flat land will depend on success in organizing tillage to follow well-established soil conservation practices that prevent erosive flow.

Cultivation on the contour

The essential rule is that all plowing and planting of row crops should follow the contour. For mild slopes on stable soil types, runoff from contour tillage may be checked by grassed strips or by narrow fields of protective cover crops such as alfalfa, stylo, or other forage legumes. This cropping pattern has been developed in the U.S.A. to highly successful systems of alternate husbandry in which contoured fields grow forage alternately with other crops. The resulting spectacular patterns are familiar to air travelers, but it is important to remember that this solution to the problems of soil erosion is valid economically only where at least half of the farmer's income is obtained from livestock. On small subsistence farms, such strips are grazed by tethered livestock.

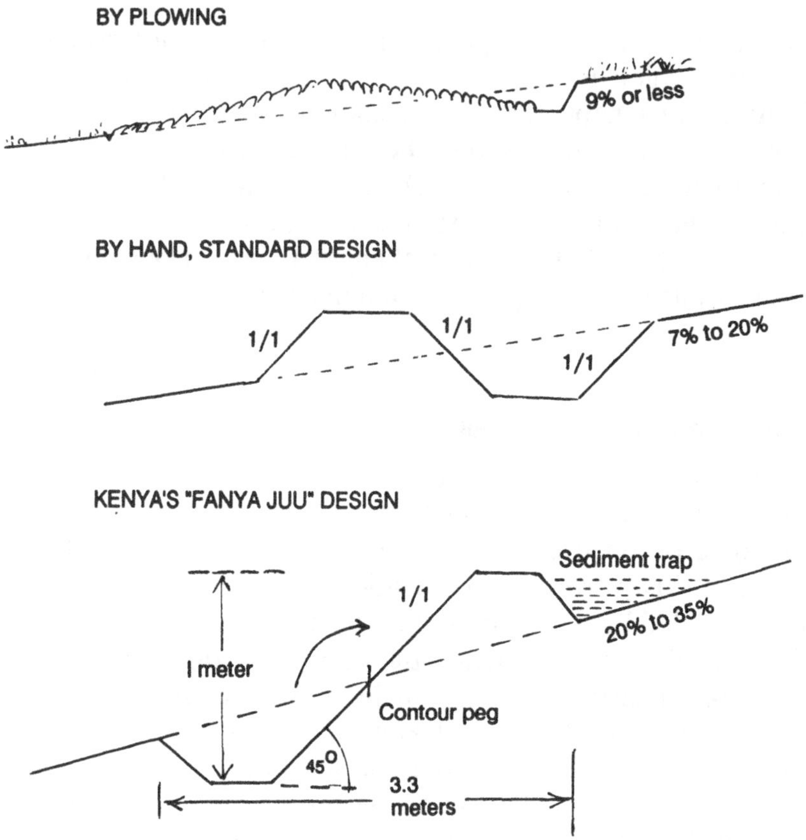

Figure 27. The dimensions of narrow-based terraces, or terrace channels, vary with the slope and soil type. The Kenya design, for manual construction, moves the soil uphill.

Graded terrace channels

For slopes up to 12 percent (1/8, or about 7°), contour cultivation needs only intermittent control of drainage by graded terrace channels, sometimes called narrow-based terraces. The term *hillside ditches* was formerly applied to obsolete patterns of deep trenches, but it is sometimes used today for graded terrace channels. These are narrow-banked channels, usually grass covered, that convey surplus water along slight gradients of about 0.25 percent (1/400) to prepared waterways (Fig. 27) or to stable natural gullies. They also provide useful paths for access in weeding or harvesting. The spacing or vertical interval (V.I.) depends primarily on the slope but is greatly affected by soil type and structural condition. Several formulae have been devised for this spacing. The U.S. Department of Agriculture formula V.I. = 0.3[2 + (% slope/4)] meters was successful in Kenya on the more erodible basement complex soils, but gave terrace channels too closely spaced for the very stable red volcanic soils.(15) In India attempts have been made to allot coefficients for soils, crop cover, and rainfall intensity.(17) In practice, the spacing must be modified through local experience. Too close a spacing was shown in experiments in Kenya to depress overall returns by reducing the area of crop without increasing the available moisture in the first 3 meters of soil depth.(15)

The Kenya studies also demonstrated that on the deep Kikuyu red loam, the runoff from contour-plowed slopes can be reduced by tied ridges. Ridge-and-furrow contour plowing gives ridges about 20 centimeters high and 60 centimeters apart. The furrows are closed, or tied, at about 1 meter intervals by small dams of soil. The ties should be about two-thirds the height of the ridges and must be made immediately after plowing because open furrows may concentrate runoff into low spots and initiate gullies. The ties are made by hand hoe or by trailing a disk held perpendicular to the furrow and lifted at intervals. Such tied ridges are an additional precaution; they do not replace graded channel terraces (Fig. 28).

Bench terraces

For control and routing of runoff on slopes steeper than 12 percent (7°), the hillside should be cut into broad flat steps known as bench terraces. When the soil is deep enough, they should be given a backslope of about 2 percent (1/50) to guide runoff to a shallow drain at the foot of the terrace bank, or riser (Fig. 29). These terrace drains should be graded gently, at 0.17 to 0.33 percent (1/600 to 1/300), in order to bring the runoff to a prepared waterway for discharge down the slope. This usually requires that the terrace be given the same longitudinal slope. On hillside slopes up to 30 percent (17°), the terrace banks, or risers, are usually given a slope of 100 percent (1/1, or 45°) unless the soil is of a type capable of standing unsupported at steeper angles. On hillside slopes steeper than 30 percent, such banks would take up too much of the crop space, so that vertical stone walls are used.

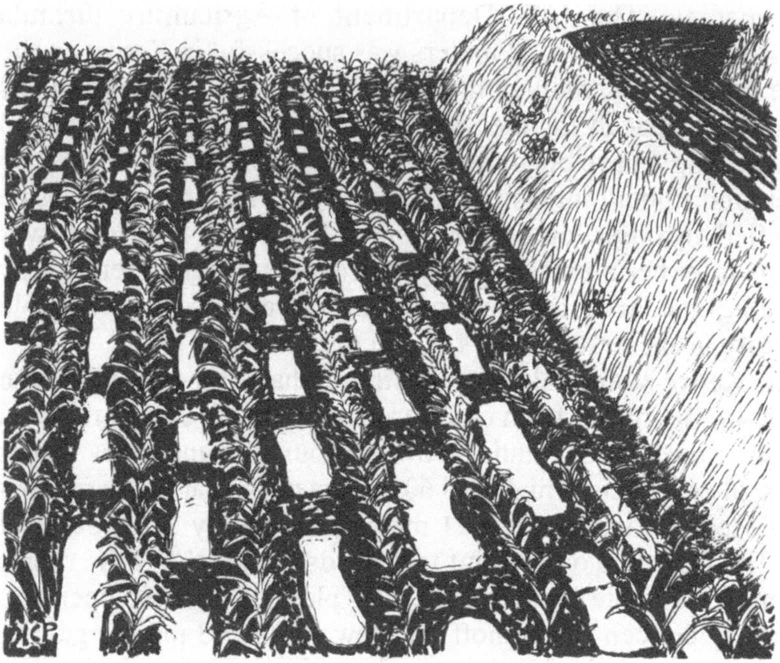

Figure 28. Under heavy tropical rainfall, even well-constructed bench terraces lose water through runoff. Use of tied ridges, also known as basin tillage, can help to retain more water for infiltration.

In some of the drier islands of Indonesia, farmers plant dense hedges of fast-growing leguminous trees to create the terrace walls. *Gliricidia maculata* and *Leucaena glauca* are used; the giant ipil ipil (*L. leucocephala*) is being tried, but its root system may prove too strong a competitor for water. The trees are cut back annually to about 2 meters high, the branches are wedged horizontally behind the trunks to reinforce the terrace walls, and the leaves are used as mulch or fodder.

Earthen banks should be planted with grasses or, better, with forage legumes such as alfalfa (lucerne) (*Medicago sativa*) or stylo (*Stylosanthes guanensis*) that have strong, deep roots. Local agricultural advisers should know the grasses most suited to terrace banks. *Eragrostis curvula* (weeping lovegrass) is much used in the U.S.A. and in high-altitude areas of the tropics. *Paspalum notatum* gives a slow-growing but very dense and mechanically strong cover. It is favored in East Africa because its

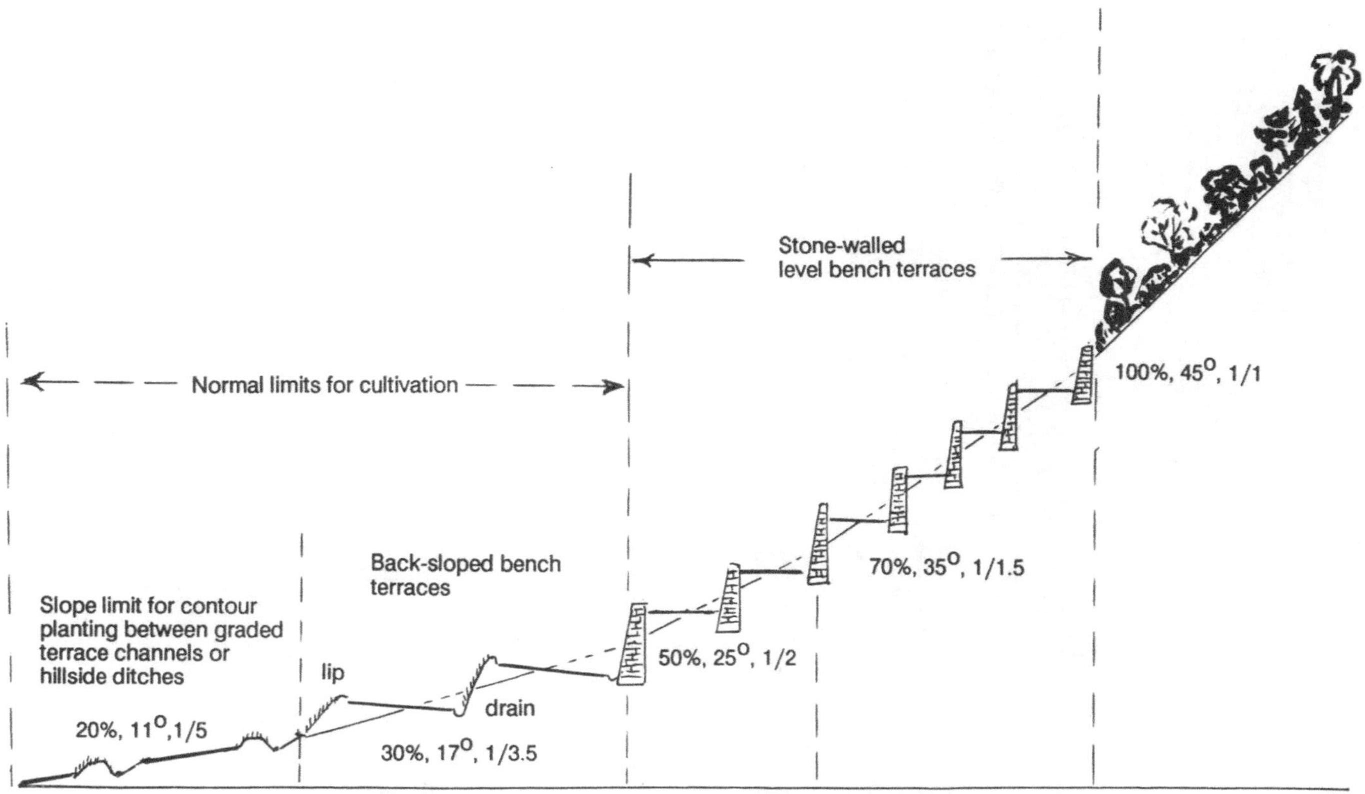

Figure 29. In tropical countries with high population densities, steep slopes are terraced by hand. With good maintenance, these terraces remain stable, but when they fail, severe damage may be caused downstream.

Stable bench terraces on a 60 percent slope in Nepal. The roadside wall is of stone, but the terrace walls are of earth.

roots, although strong and dense, do not penetrate far into the crop root zone. Experience in Indonesia favors *Brachiaria brizantha* as the most suitable grass for terrace banks.(1) Where fuel is in short supply, the terrace banks can be planted with fast-growing trees (see Chapter 5).

The labor invested by the farmer in bench terrace construction is very substantial. For building by hand, 600 to 1000 laborer days per hectare are needed. It is therefore critically important that the terraces be properly designed and built to last. Those that have lasted for many centuries (see Chapter 2) have usually been built for irrigation. Their intrinsic advantage is that they have to be leveled accurately in order to function. They have a lip on the outer edge and some form of drain and spillway to dispose of surplus water safely. They are therefore able to survive heavy rainfall. A narrow lip or soil bund along the outer edge is a valuable addition to rainfed terraces because it protects the terrace edge from damage by draft animals, it stops stormwater from spilling over and starting a gully, and it prevents the planting of a row of crop whose harvesting will damage the edge. In Java, I have observed terraces that are frequently and unnecessarily damaged by harvest of cassava planted to the edge.

A critical aspect of terrace construction is the retention of the topsoil. With half of the terrace in "cut" and the other half in "fill," it is easy to bury fertile topsoil under subsoil of poor structure and low nutrient quality. This can reduce crop yields for several years. Where the soil is not deep enough for backsloping terraces, they may be built level or with gentle forward slopes. If the forward slope is only 2 or 3 percent and is within the stability limits of the soil, the structure may be durable. However, the spread of cultivation onto steep slopes with thin soils has produced dangerously ineffective forms of outward-sloping terraces.

Narrow terraces for tree planting

On slopes steeper than 100 percent (45° or 1/1), the area gained for cropping becomes too small to repay the labor of constructing and maintaining terraces. It may, however, be necessary to stabilize such steep slopes by reforestation where firewood cutting

Well-built stone terrace walls in Jamaica. (Photo: G.K. Argles)

Live hedgerows used to support terraces in the dry northern coast of Bali, Indonesia.

In studies of legumes for terrace banks, none has yet been found to compete with the *Brachiaria*.

and overgrazing have caused sheet erosion. Trees are then planted on narrow ledges, hand cut on an horizontal alignment, with a 10 percent backslope. These tree-planting terraces are only 0.5 to 1.5 meters wide. This manual labor is undertaken on an impressive scale in northwest China.

Where trees have already been planted in straight lines that ignore contours, as in some early coffee plantations, and soil erosion is becoming evident, a practical remedy is to cut narrow-terrace channels through the plantation using a plow or a small grading blade. Surprisingly few trees have to be removed from a normally spaced (2.8×2.8 meter) stand. The resulting control of erosion can lengthen the life and increase the productivity of the plantation.

Individual tree terraces are used for hillside orchards with a grass or legume ground cover, broken by separate semicircular platforms, arranged in staggered rows or "fish scale" formation.(16) This system requires far less soil shifting than construction of continuous terraces. It has been used in Malaysia for planting oil palm on steep hillsides.

Terrace banks in Java stabilized by planting a vigorous grass, *Brachiaria brizantha.*

Waterways and drop structures

In order to shorten the length of flow from terrace to waterway, it is usual to divide all types of graded terrace by starting at a high point about midway between waterways and grading outward (Fig. 30). Leading the runoff from terraces of all types along gently sloping drains to the disposal channels requires only simple earthen ditches.

Getting the accumulated water safely down a steep hillside requires much more construction. Some hillsides are dissected by deep natural gullies. If these are stable, because they are lined with large rocks and boulders or because the stream has cut down to bedrock, they are the optimum routes for the disposal of terrace runoff, which is led into them. Where no such easy solutions are available, open channels with low banks are dug and planted with grass stolons or sown with grass seed to obtain a complete cover before they are brought into use. They will then remain stable and carry water at speeds of up to 2 m/sec without damage down slopes of 12 percent (7°) or less. For steeper slopes, the water must be made to lose height and energy by small drop structures of stone, timber, brushwood, or bamboo. Stone structures are best because wooden structures need more maintenance. For stone-walled terraces, a notch is made in the walls. The rapid undercutting that will result from the scouring action of the falling water should be prevented with a protective apron of stones or wood. Stronger drop structures are needed near the bottom of the slope where the accumulated water volumes must be controlled. Gabions of wire mesh filled with stones are easier to build at remote sites than are concrete structures (Fig. 31). Careful linking with roadside ditches and culverts is necessary to integrate the watershed drainage system.

Diversion, or cutoff, drains

Many terraced areas in Nepal, in which much labor has been invested, have been destroyed by severe gullying from misused hill land above them (see Chapter 7). This illustrates the great importance of protecting terrace systems from the runoff generated by higher land. The protection is secured by digging storm-

Tree-planting terraces on steep slopes in China's loess hills.

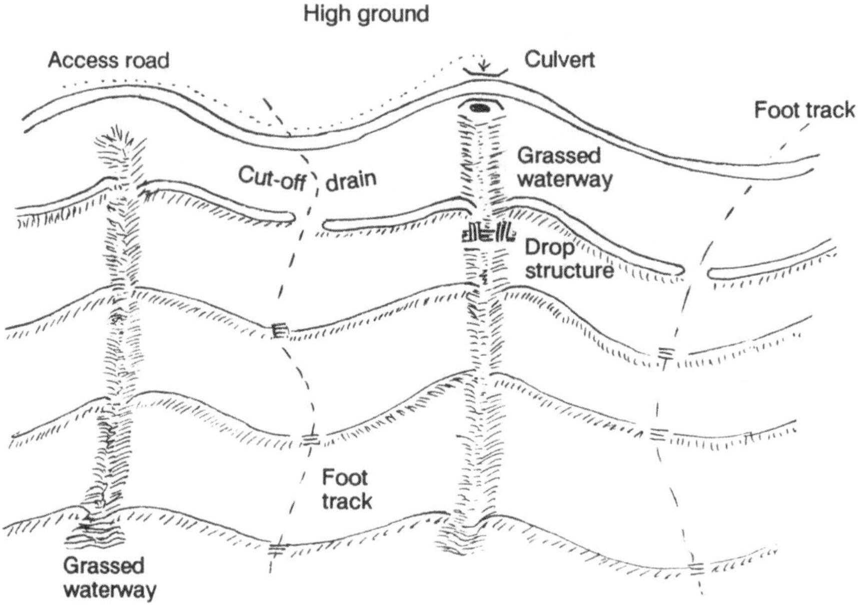

Figure 30. Foot tracks follow the ridges. Where they cross terrace banks, steps should be cut and brushwood pegged across the track. Where road culverts discharge into grassed waterways, drop structures are usually needed.

water-diversion, or cutoff, drains to trap the runoff and convey it safely to a natural drainage line.

The dimensions of the drain must depend on an estimate of the volume of water to be diverted. It is best dug with sloping sides, rather than as a rectangular section, with the soil banked on the downhill side to increase the capacity (Fig. 32) The discharge of such diversion drains should be into a stable natural channel, if possible over rock, because, as for road culverts, the diversion drain itself could become the source of a gully.

Figure 32 is a field sketch of a real situation in northern Kenya. A gully was cutting back several meters in each rainy season, as it received the drainage from a flat and shallow basin entirely hidden under a flourishing crop of maize. A simple diversion drain to a well-vegetated waterway cured the trouble. In addition to constructing the diversion drain, it was necessary to lead water away from the flanks of the actively growing gully by digging graded narrow channel terraces. All of the necessary action was within the means and abilities of the subsistence farmer, but he needed guidance and confidence in the outcome before undertaking the labor. A vast amount of tropical soil erosion is due to lack of awareness and to a philosophical belief that nature has clearly ordained soil and water to run downhill.

Water harvesting

An extreme case of runoff control is found at the limits of farming in the semiarid zones. This is the routing of runoff from a large area of thin-soiled or rocky hillside down to a small area of deep soil (as described in Chapter 6). Along the Mediterranean coast of North Africa, runoff has traditionally been stored in the alluvial deposits in the beds of wadis, which flow only a few times each year, usually as torrents. The alluvium is stabilized by building a series of low stone walls about 1 meter high, which form a "staircase." These small dams trap sediment and can fill in a single season. Part of the floodwater is stored in the fill. The water is available to crops, usually barley, that are sown on the fill.

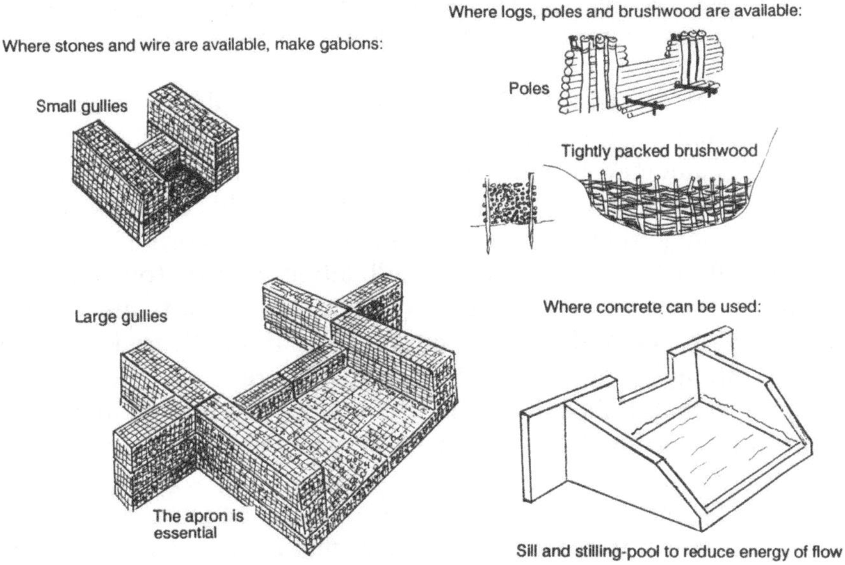

Figure 31. Gully-stopping structures. Gabions are wire mesh cages, usually woven on the site, that are filled with stones or broken rock.

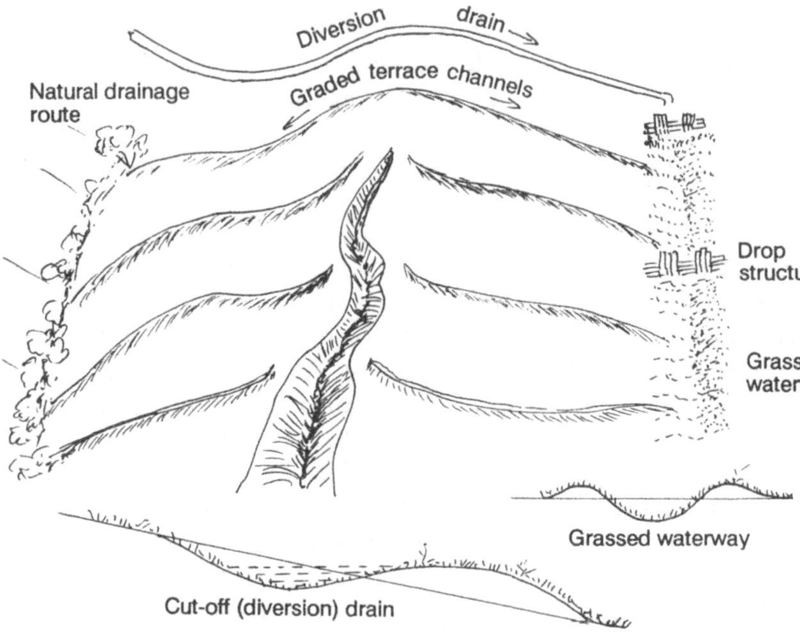

Figure 32. Cutoff drains (stormwater diversion ditches). In the treatment of an actively eroding gully, the cutoff drain diverts water from the head, while the terrace drains lead water way from the sides. Stabilization can be accelerated by planting fodder grasses and fuelwood trees in the gully.

Where such wadis discharge onto dry alluvial flats, there is a cone of coarse gravel and sand into which much of the water sinks. An ancient Iranian tradition has been to sink wells 10 meters or more into such water-bearing gravels and then to dig gently sloping underground tunnels, or *qanat*s, to bring the water to the surface. The cost in skilled labor is now too great to maintain these ancient systems; instead well points are driven from the surface and water is pumped up and delivered through plastic piping laid on or just below the surface.

Modern materials can now improve many traditional methods. Even the traditional hand-built stone walls across wadis can be constructed more strongly by using gabions of wire mesh (see Chapter 2). The retention of silt and sand can be improved by the use of woven nylon fabrics that are manufactured for soil stabilization in engineering constructions. Other methods of harvesting runoff are given by Frasier.(5)

Constructional measures to check overland flow have proved effective where land misuse and excessive runoff have lowered water tables and so increased desertification. In the Quetta region of Baluchistan under 200 millimeters of annual rainfall, a project supported by UNDP/FAO built low dry-stone walls across drainage lines and dug shallow contour trenches across the lower hill slopes. These structures increased the infiltration so effectively that the groundwater levels in the valley rose by 10 meters. Unfortunately, uncontrolled borehole drilling by private landowners lowered the groundwater levels again, but the principle was well demonstrated. Runoff from these hills is also used to improve pastures.(6)

In semiarid cropland where yields are frequently limited by water stress, it is advantageous to lead the runoff into small storage ponds for use as supplementary irrigation. The International Crops Research Institute for the Semi-Arid Tropics has developed good evidence in India that single applications of 50 millimeters of water at critical periods can increase crop yield substantially: in sorghum, to 3.9 t/ha from 1.9 tha; in millet, to 3.2 t/ha from 2.0 tha; and in maize, to 5.7 t/ha from 3.0 t/ha in well-fertilized crops on red soils (alfisols) and to 2.4 t/ha from 1.6 t/ha on black cracking clay soils (vertisols).(11) Storage ponds are also valuable for watering tethered or stall-fed livestock in the croplands.

Setting out

An essential service to hillside subsistence farmers is help from trained agricultural staff in the design of terraces and in setting out the pegs for construction. This work does not necessarily require expensive survey instruments such as theodolites or quick-set levels, although the setting out can be done more quickly and more accurately with them. Instead, slopes may be measured with the simple hand-held Abney level, and contours can be set out by the Kenya-type line level, which is essentially a carpenter's spirit level suspended from two sticks in the middle of 10 meters of cord (Fig. 33). For level terraces, the sticks are of equal length; for a drainage grade along a terrace of, say, 4 percent, one stick would be 25 millimeters shorter than the other.(13) A simple Australian field device is a length of water-

Figure 33. Use of the line level.

filled hose pipe with glass observation tubes at each end. Setting out by eye only is hazardous, except for irrigation terraces, where the flow of water immediately demonstrates any errors. The danger of badly leveled terraces is that the concentration of overland flow to the low spots can initiate gullies. A poorly made terrace can cause more damage than the untreated soil erosion from the slope.

The details given here are intended to explain the problems and describe practical solutions. For those who have technical responsibilities in the field for soil conservation, engineering handbooks such as those by Hudson(9, 10) and Singh et al.(17) are essential.

Community organization for watershed discipline

For maximum effect, the watershed plan for roads, farms, terraces, storage ponds, and stockpens should be drawn up in advance of settlement. In most areas of tropical countries, however, the farmers are already settled, and the main need is to correct the misuse of land at various stages of damage. Watershed boundaries rarely coincide with either farm boundaries or administrative divisions. The safe routing of runoff water, however, depends on the shape of the land and the location of the natural drainage lines; these lines may cross both farm and district boundaries. Cooperation of all farmers in a subwatershed is therefore essential.

The initial subwatershed planning ideas must be presented, by someone with the necessary training, for discussion with all holders of land titles or rights of use. When the majority is persuaded that the control of soil erosion is advantageous, a detailed plan can be made. The person who prepares the plan should have technical training but not necessarily at the university graduate level. The work should be supervised by an agricultural engineer or agronomist specializing in soil and water management who should give advice on site rather than from an office.

Waterways are the most critical issue. They occupy little land area, but where family holdings are only half a hectare, land is precious. The grassed waterways should produce good forage, and each farmer should cut the grass on his or her own land, but there must be some agreed community sanction against digging or plowing across waterways. Such reorganization and the heavy labor entailed require a combined effort by all farmers in the watershed. Equally important is technical and financial help from governments, acting on behalf of the larger communities lower in the valley that will either gain from cleaner and more regulated waterflows or suffer from the soil degradation of the upper watersheds.

Intensive conservation areas

The concept of community organization for each subwatershed of 20 to 100 farmers or more was developed in the U.S.A., with the backing of legislation. There the law takes effect only when the plans are accepted by a large majority of the farmers and right-holders involved, but it then becomes enforceable on any remaining objectors. The farmers elect their own committees and chair them; an extension agent gives advice at the meetings.

This procedure has been found successful for African small farm-holders in Kenya and in Zimbabwe, among those growing cash crops. It is less easy to introduce in the more strongly tribal subsistence areas, but the tribes have their own forms of community organization, which have proved highly effective in group efforts for terrace construction. It is, however, unconvincing to such communities unless soil conservation is part of a package to increase yields with better varieties and by use of fertilizers where supplies of manure are inadequate. Unless yields improve, terraces will be neglected.

For the forested upper section of the watershed, a successful communal organization of forest management in Nepal was described in the previous chapter. Here the people living high in the middle mountains have responded positively to recent legislation permitting elected village councils (panchayats) to manage an area of their watershed forests. Progress depended on the availability of a district forest officer to give technical guidance and encouragement.

An example from China

Rural community organization for soil conservation and reforestation of hills stripped naked by firewood cutting is making remarkable progress in the loess areas of China. The earlier feats of the brigade system in land reshaping and replanting have been followed by a major new impetus from the "individual responsibility law" of 1981. Under the law, peasant farmers who volunteer to terrace the hills are offered inheritable titles to land that they have reclaimed as crop terraces or to steep hillsides that they have reforested. A small government subsidy helps to maintain the family during this long and heavy labor. For very steep hillsides with more than 100 percent slope (45°) on which narrow terraces are cut for tree planting, one-half the subsidy is withheld for a year, and the farmer is paid only when an 85 percent stand is achieved. High capacity nurseries of the Forest Department supply the trees.

An example from Java

Java, a fertile volcanic island with annual rainfall ranging from 1500 to 3000 millimeters, carries an astonishing density of population, with a rural average of some 700 people per square kilometer. The maintenance of soil fertility and the care of topsoil are therefore of critical importance. An intensive campaign of soil conservation, supported by USAID and UNDP/FAO, is based on a series of demonstration areas, or model farm areas, each about 10 hectares, that involves a group of about 25 key farmers. The model farm is surrounded by an extension area of about 100 hectares in which farmers are subsidized to follow the demonstrated practices. There is a substantial payment (about U.S. $200/ha) for the model area for the first year only. Smaller 2-year subsidies are offered in the extension area. The main changes are improved bench terraces, with a lip on the terrace edge, protected by planted grass or legumes on the risers. The most successful grass is *Brachiaria brizantha;* compatible legumes are needed that will survive the competition from the grasses.(1)

Credit schemes are being introduced, with external support, for purchasing seed, fertilizers, insecticides, and goats. The idea of providing credit for goats would seem incredible to a soil conservation officer from Africa, where the misuse of the hardy goat in communal grazing is an acute problem. However, in Java, the goats are stall-fed or tethered because a general consensus of opinion does not tolerate free-ranging animals. The farmers will more readily plant grass on their terrace banks if they can make profitable use of it. Credit for goats is thus a sound policy. The administration of this major agricultural program is extremely complex, with several ministries represented at district level. The results I saw in 1984 were excellent in the more prosperous areas, but progress was slower in the areas of apparent poverty, where land tenure and sharecropping problems inhibited response.

Community track discipline

Life in rural tropical villages involves much walking. Footpaths in hill areas of high rainfall can become serious sources of gully erosion, especially in deep-volcanic or loess soils. If they are also used by livestock, the damage is greatly increased. Footpaths therefore need careful attention in soil conservation plans. They should follow terraces to the middle ridge or line of high spots at which the graded channels diverge and then follow these ridges uphill or downhill. Where slopes are steep, simple steps made by driving in pegs and laying branches behind them

A footpath in an Ethiopian village that will cause a gully through the terraces. It is important for communities to plan and regulate formation of footpaths.

can both protect the soil and assist the walkers. It is particularly important that footpaths should not cut through the lips of terraces, because footpaths can rapidly become drains (see Fig. 30).

References

1. Barrau, E.M. 1984. *Production constraints and soil erosion in the humid tropics of densely populated Java.* Jakarta: U.S. Agency for International Development.
2. Duley, F.L. 1939. Surface factors affecting the rate of intake of water by soils. *Soil Science Society of America Proceedings* 4:60.
3. Ekern, P.C. 1950. Raindrop impact as the force initiating soils erosion. *Soil Science of America Proceedings* 15:7–10.
4. Ellison, W.D. 1944. Studies of raindrop erosion. *Agricultural Engineering* 25:131–136, 181–182.
5. Frasier, G.W. 1984. Water harvesting including new techniques of maximizing rainfall use in semi-arid areas. *Proceedings of the fourth agricultural sector symposium.* Washington D.C.: World Bank.

6. French, N., and Hussain, I. 1964. *Water spreading manual.* Range Management Record no. 1. Lahore: West Pakistan Range Improvement Scheme.
7. Grant, P.M. 1967. *Annual Report.* Salisbury, Rhodesia: Agricultural Research Council of Central Africa.
8. Hudson, N.W. 1957. Erosion control research: Progress report on experiments at Henderson Research Station. *Rhodesian Agricultural Journal* 54:297–323.
9. Hudson, N.W. 1975. *Field engineering for agricultural development.* Oxford: Clarendon Press.
10. Hudson, N.W. 1981. *Soil conservation.* London: Batsford Academic.
11. Krantz, B.A.; Kampen, J.; Russell, M.B.; Thierstein, G.E.; Virmani, S.M.; and Willey, R.W. 1982. *The farming systems research program.* Hyderabad: International Crops Research Institute for the Semi-Arid Tropics.
12. Laws, J.O. 1940. Recent studies in raindrops and erosion. *Agricultural Engineering* 21:431.
13. Layzell, D. 1968. *Construction and use of the line level.* Nairobi: Kenya Department of Agriculture.
14. Pereira, H.C.; Chenery, E.M.; and Mills, W.R. 1954. The transient effects of grasses on the structure of tropical soils. *Empire Journal of Experimental Agriculture* 22:148–160.
15. Pereira, H.C.; Hosegood, P.H.; and Dagg, M. 1966. Effects of tied ridges, terraces and grass leys on a lateritic soil in Kenya. *Experimental Agriculture* 3:89–98.
16. Sheng, T.C. 1977. Protection of cultivated slopes— terracing steep slopes in humid regions. In *Guidelines for watershed management.* Conservation Guide no. 1. Rome: FAO.
17. Singh, Gurmel; Venkataramanan, C.; and Sastry, G. 1981. *Manual of soil and water conservation practices in India.* ICAR Bulletin T.13/D.10. Dehra Dun: Indian Council of Agricultural Research.

9
Management problems of alluvial lowlands

Assets and hazards

As in upper watersheds, detailed land-use planning and management are required in the lowland areas of river basins. Their wide ecological range includes the hot, dry coastal plains of East Africa and the hot, wet forested lowlands of the Amazon basin, both of which support only sparse populations. However, some of the irrigated flat alluvial floodplains, such as those of the Indus and the Ganges, carry hundreds of millions of people. The management of the dry rangelands has been discussed in Chapter 7 and that of the wet forests in Chapter 4. This discussion will concern the management of the lowland alluvial areas where watershed policy is of major concern.

Tropical lowland alluvial areas often combine cultivatable soils, accessible water supplies, and year-round growing temperatures. But they also present formidable management problems. The foremost is the hazard of floods, which is not surprising because floods were the geological mechanism by which these plains were built. Also, in the tropics and subtropics, the prevailing high evaporation rates make salinity a pervasive problem. Although technical solutions to most management problems of alluvial lowlands are now known, land affected by salinity is still the largest remaining undeveloped food resource in the Third World.(11)

Potential for irrigation

In order to sustain crop productivity, commercial farmers throughout the world use fertilizers, but most subsistence farmers of the tropics do not yet do so. This is in part because they lack access to both supply and credit and in part because of the uncertainties of rainfall or drought under the ITCZ. The alluvial plains provide both easier supply of inputs by road, rail, or water transport and the critical opportunity for irrigation from rivers or groundwater. Management skills remain as acute a constraint in the lowlands as in the highlands so that in spite of these advantages, average yields of irrigated crops in developing countries are disappointingly low. Figure 34 demonstrates the startling range of yields, on a whole-country basis, for rice, the world's main irrigated crop.

In a survey for the U.N. World Water Conference in 1977, FAO reported a world total of 223 million hectares of equipped irrigation areas, of which 92 million were in developing countries. Nearly half of the latter, some 45 million hectares, were reported to be in need of reclamation and rehabilitation. This figure indicates the immensity of problems in managing lowland tropical soil and water resources. Two major sources of improvement are already having an effect on a large scale. One is the development of high-yielding varieties, which repay profitably the supply of plant nutrients; the second is the improvement of on-farm water distribution and drainage systems.

The Green Revolution

In the 1930s, attempts to improve yields of tropical grain crops, principally rice, sorghum, maize, millet, and some wheat, were disappointing because the strains grown by subsistence farmers had been selected during the course of centuries to survive at low levels of nutrient supply. They grew slowly, produced one crop a year, and failed to respond to addition of fertilizer. In the 1950s, research teams of the Rockefeller Foundation working in Mexico on wheat and maize broke through this constraint in wheat by a combination of plant breeding and improved agronomy. Mexico progressed from a grain-importing to a grain-exporting country, a situation that was later reversed by population growth.

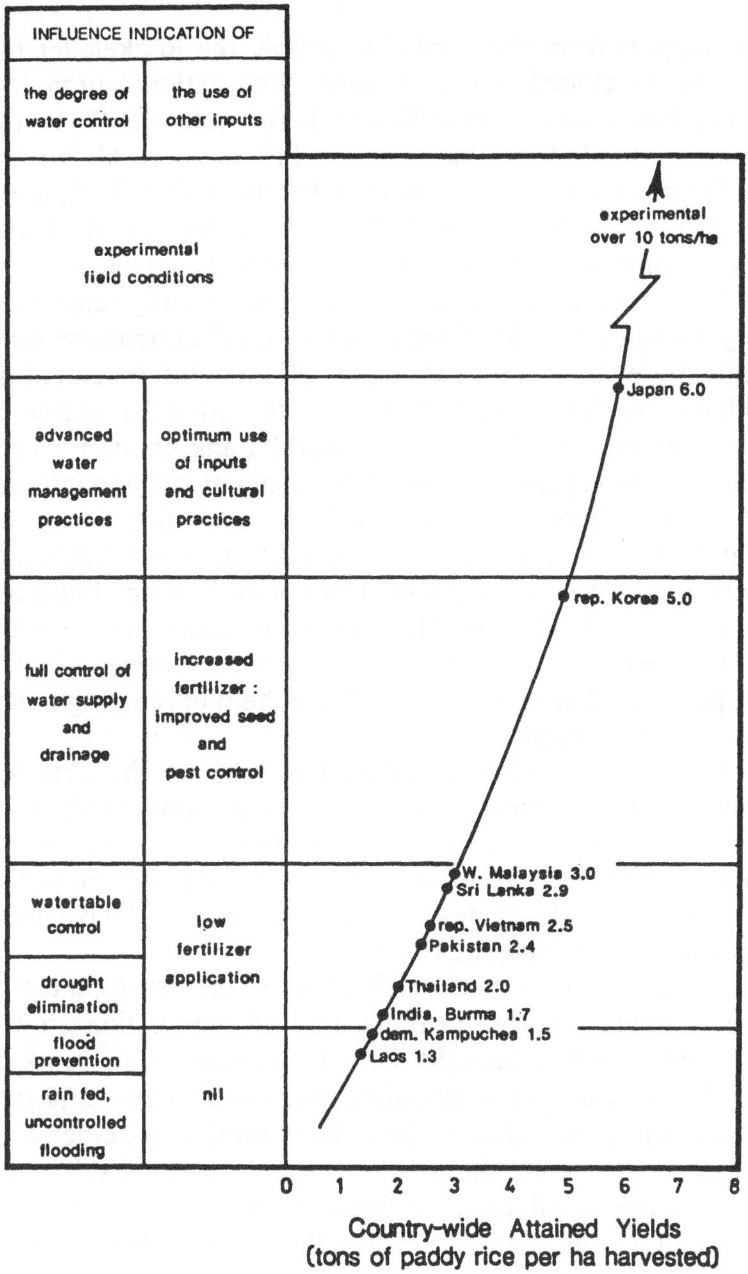

Figure 34. The world range of rice yields. The International Rice Research Institute has demonstrated that very great improvements are possible in low-yield countries through cooperation with local scientists. *Source:* FAO, 1979, *The on-farm use of water.* Rome.

With support from the Ford Foundation, the Rockefeller initiative was developed into two strong international organizations, the International Rice Research Institute (IRRI) in the Philippines, founded in 1960, and the International Maize and Wheat Improvement Center (known by its initials in Spanish as CIMMYT), established in 1967. Both centers pursued vigorous programs of plant breeding and agronomy to produce a range of modern fertilizer-responsive varieties usually called high yielding varieties (HYV). Centers for tropical agriculture were subsequently established in Colombia (Centro Internacional de Agricultura Tropical) and in Nigeria (International Institute of Tropical Agriculture). In 1971, the World Bank chaired a consortium of donors, private foundations, and national and international aid organizations and set up the Consultative Group for International Agricultural Research (CGIAR), with a Technical Advisory Committee of scientists from both developed and developing countries. The CGIAR system now comprises 13 international centers, which operate on a slender budget of $180 million per year. The main objective is to help developing countries to grow food crops.

The major impact of this system has been on the irrigated lowlands. HYVs of wheat, rice, and maize are now used on so large a scale that the annual world increase in production is variously estimated at 30 to 70 million tons. There is no doubt that without HYVs many developing countries would have faced widespread famine in recent years.

Perhaps the most dramatic early effect in the development of modern varieties occurred in India. In 1966, population growth and poor weather had brought the subcontinent to the brink of famine. By a courageous decision, the government imported 20,000 tons of the new dwarf wheats from Mexico, together with extra fertilizers to grow them. A key step was to subsidize the fertilizer so that 1 kilogram of fertilizer nitrogen cost only as much as 3 kilograms of grain. Because the new wheats responded with a 25-kilogram increase in grain yield for each kilogram of nitrogen applied, the farmers rapidly took up the opportunity for profit.(6) Adoption began with the large (40-hectare) irrigated farms of the Punjab, but smaller farms followed rapidly. Now more than three-quarters of all of the wheat grown in India are modern varieties.

The imported varieties did not prove susceptible to local strains of plant diseases, but they have since been safeguarded by crossing with local wheats that are resistant to local pathogens. A misunderstanding about HYVs, which has appeared frequently in published criticisms of the Green Revolution, is the claim that they cannot be grown without fertilizer. There is ample evidence that they also yield better than unimproved varieties when grown without fertilizer because of their rapid growth, which reduces susceptibility to erratic rainfall patterns, and because of their broad resistance to pests and diseases.

A practical example from the Philippines

IRRI has produced fast-growing dwarf rices so that in the year-round tropical growing season three crops a year could be harvested without overloading threshing, transport, and storage facilities. Foreign voluntary workers attached to IRRI were stationed in small villages where each helped a group of 10 subsistence farmers on 0.5 hectare farms and supplied the new seeds, fertilizers, and crop production chemicals on credit. I visited three of these groups in 1975 in a district where the typical farmer produced a single crop of 1.5 t/ha, with 2 t/ha considered a very good crop. When I arrived, the groups had already harvested two successive 5-t/ha crops. But festival time then began and the farmers celebrated their astonishing yields so heartily that the third crop went in 3 weeks late and was expected to yield only 2.5 t/ha. When asked how they liked the quality of their new rice, they all replied that it was too valuable to eat. Their neighbors wanted the seed and offered two bags of local rice for one of the new IR30 variety. The yield in terms of local rice was thus effectively 25 t/ha in this first season, or about 16 times the usual yield. For them, the term *Green Revolution* was no exaggeration.

Thus, a major improvement in lowland management has been to raise the standards of irrigated farming. But yields of irrigated rice remain poor in many developing countries. The reason for this—the low standard of management of soil and water resources—is a second major constraint. The specific problems are poor distribution of water on and among farms and the very serious lack of constructed drainage channels.

Distribution beyond the canal system

A traditional view held by irrigation engineers is that their role should be to design, build, and maintain canal systems. They assume that, after centuries of practice, the tropical subsistence farmer knows best how to handle soil and water management beyond the outlets of the minor canals. In practice this is quite untrue and the lack of technical guidance to village communities on the distribution of canal water among and within farms has caused a major waste of productive resources. An economic study by IRRI of the ricelands of the Philippines showed that a major constraint on crop yields was inefficient distribution of water between farms. I have seen astonishing neglect of irrigation ditches in Pakistan, India, and Egypt. Where competent investigations have been made, measured losses of 55 percent or more have been recorded in those countries.

Watercourse improvement in Pakistan

The Indus Plain, which has the world's largest contiguous area of irrigated land, receives 120 billion cubic meters of water per annum by diversion from the Indus River system. This is double the discharge of the Nile River at the Aswan High Dam. Less than 50 percent of this great flow reaches the crops. Nearly 25 percent seeps or evaporates from the unlined canals (although some seepage is recovered by pumping from tube wells where groundwater salinity permits). Of the amount discharged from the outlets of the minor canals, 55 percent is lost in the village watercourses that distribute water among the farms.(13) Thus, a flow of water equivalent to the full annual flow of the Nile is lost between diversion from the river system and application to crops.

The major losses occur in the communally owned watercourses among the small farms. In Pakistan, the ditches are wide, shallow, weedy, and overgrown by rushes, bushes, and some trees. Water buffalo wallow in them and village laundresses trample them. The watercourses are contained by earthen banks whose integrity has been destroyed by farmers who cut outlets into fields and roughly plug them with clay, stones, brushwood, and grass.

An effective solution has been evolved at the Mona Experimental Reclamation Project in Pakistan. Supported by the U.S Agency for International Development, a team from Colorado State University (U.S.A.) conducted studies for more than 15 years with watercourse construction and linings of brick, concrete, and various plastic sheets. The least-cost solution was reconstruction of the earthen bunds with hand tools to form a flat-bottomed channel section (Fig. 35) with thorough compaction. Concrete is used only at the outlets to each farmer's fields, at junctions of water courses, at special bathing tanks for buffalo (Fig. 36), and at platforms for village laundries. The resulting channel halves the water losses. The water tables are lowered, and crops visibly benefit from less waterlogging and salinity. The ratio of benefits to costs ranged from 6:1 to 2:1.(8) Farmers have responded well, contributing labor for reconstruction and maintenance while government supplies supervision and cement. The

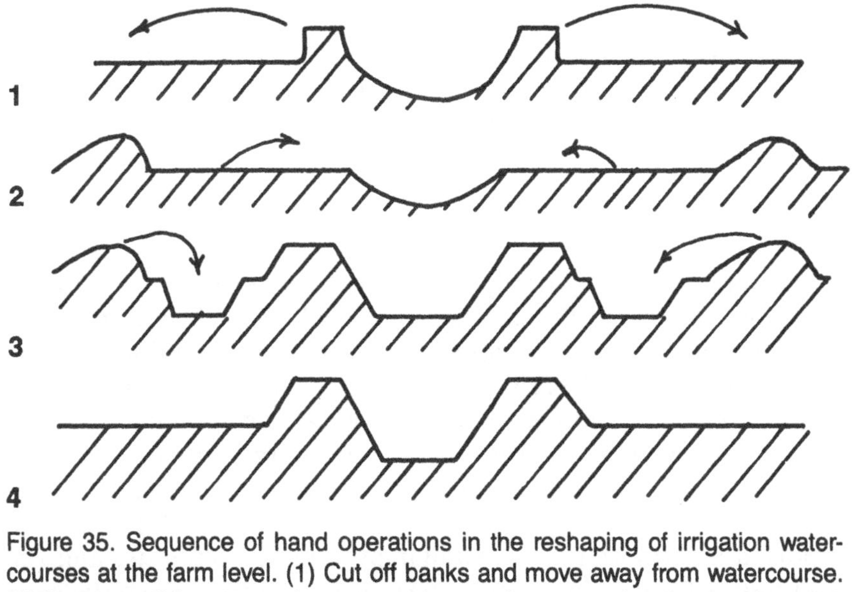

Figure 35. Sequence of hand operations in the reshaping of irrigation watercourses at the farm level. (1) Cut off banks and move away from watercourse. (2) Borrow soil from near the watercourse and compact into layers about 7.5 centimeters thick to form new banks. (3) Pull soil from old banks into the borrow pit. (4) Reshape watercourse. *Source:* Mona Reclamation Experiment Project, WAPDA, Pakistan.

Figure 36. Water buffalo should be provided with wallows reinforced with stone or concrete. Otherwise, earthen irrigation channels will be damaged by these heat-sensitive animals.

water saved is important, and the long-term plans for irrigation development will rely on this reduction of losses for 40 percent of the additional irrigation water.(14)

The scale of the problem is immense. In Pakistan, the USAID project has extended the improvement to some 1500 villages, and projects of the World Bank, International Development Association, and International Fund for Agricultural Development have trebled this target to 4600 villages, but there are more than 89,000 villages needing attention.

Watercourse improvement in India

An approach comparable to the work at the Mona station has been undertaken in India, where the problems of on-farm water distribution are a serious constraint on crop production in many irrigation command areas. Replacing the traditional inefficient field-to-field irrigation with well-designed earthen channels permitted improved crop varieties to be profitably grown with fertilizers. Various studies have found high benefit-cost ratios. For example, Easter (2) reported a ratio of 10:1 and Kumar (9) reported a ratio of 16.5:1. Similar problems, solutions, and opportunities for improvement of farm water courses are to be seen in Egypt, Thailand, and the Philippines; in China, however, a traditional care for water courses prevails.

Drainage and salinity

Both river water and groundwater contain salts dissolved from soils and rocks in the upper watersheds. Salts rarely cause difficulties until they reach the lowlands, where they are concentrated by evaporation. If the surface of the groundwater, or water table, is high enough for soil moisture to reach the surface by capillary action, the salts may accumulate in the surface soil at concentrations that are lethal to all but specially adapted plant species.

The remedy is to improve the drainage until the water table is 3 meters or more below the surface. The salts can then be leached away by liberal applications of water. Salts, however, continue to arrive in the irrigation water and to be drawn to the surface as water evaporates from the soil, which is repeatedly wetted by irrigation. Moreover, excess water for leaching is not available in years of low riverflow. Management must therefore be continually vigilant.

Egypt has the world's largest program for field-piped drainage. With international funding, several groups of contractors are using the latest laser-beam leveling techniques to lay plastic drainage pipes under more than 2 million hectares of irrigated croplands. The work goes on right through the crop season, and I have been intrigued to observe the eager cooperation of farmers. Agents travel ahead of the tractors and buy the standing crops at just enough above market prices to ensure a ready welcome. Watershed improvement schemes involving peasant-farming communities can succeed or fail on such small details. Measured crop yield increases from earlier trials have ranged from 20 to 30 percent for complete districts and up to 50 percent for groups of farmers linked by a main drainage channel.(3)

Drainage is similarly recognized as the major management problem in India and Pakistan. A major drainage channel has been built in Pakistan, with World Bank support, parallel to the western bank of the Indus to carry saline water to the sea. Construction of a similar drainage outfall to the sea, some 192 kilometers long, on the eastern bank, began in 1986.

Iraq: An ancient problem and a modern solution

The earliest historical records of irrigation management indicate that the problem of salinity and its solution were well known. Large-scale irrigation systems were constructed by successive regimes occupying the floodplains between the Euphrates and Tigris rivers from about 2400 B.C. onward. Irrigation was repeatedly interrupted by warfare. Sumerian inscriptions on clay tiles record the control of salinity by leaching from canals fed from the Euphrates and discharging into the Tigris.(7) Early in the 3rd century A.D., the Persians constructed a brick diversion weir across the Tigris near Samarra. Water was delivered via the Nahrawan Canal, some 140 meters wide and 300 kilometers long, to irrigate the desert on the eastern bank. Prosperity depended on authority to maintain canals against annual sedimentation and damage by large floods and to control salinity by leaching. Disruption of such authority through repeated warfare caused the collapse of the main canal systems by about A.D.1600. In the 1950s, flood-control dams and new canals were built to regain command of about 4 million hectares, but management of irrigation and drainage was not effectively restored. The crop root zone has been restricted by rising water tables and increasing salinity, so that less than one-half of the area is cropped. The yields are low; the villages are wretchedly poor.

With international help, the government of Iraq has made impressive progress on a vast new drainage and irrigation program, which aims to cover 3.5 million hectares. A main drain, referred to as the Third River, runs between the Euphrates and the Tigris from north of Baghdad to the sea, fed at 1-kilometer intervals by open drainage channels up to 3 meters deep. The plans require the digging of 2000 kilometers of these drainage channels annually for the next 25 years. It is sad that this important work has been interrupted by war.

Drainage by tubewells

Estimates of the losses of productive irrigation area to waterlogging and salinity throughout South and Southeast Asia during the 1950s indicated that these probably exceeded the rates at which new areas were being commanded and equipped. This trend has fortunately been reversed in the last 3 decades by the great increase in the use of borehole pumps, following the rapid extension of electrical power networks. Where the groundwater is fresh (that is, nonsaline), it represents an immediate water resource that can be developed at far less cost than canal irrigation. Pumping lowers the water table until the capillary fringe of soil moisture, which climbs to about 1.5 meters above the free water level, no longer reaches the surface, so that salts no longer accumulate in the surface soil and past accumulations are leached downward.

Where the groundwater is saline, the levels can be usefully reduced by pumping, but the water must be disposed of by a drainage system. This technique has reached a high level of development in Australia's Murray River alluvial lands. There saline water is pumped to large natural depressions or salt pans for evaporation, rather than into the Murray River where it would harm the water supplies for the city of Adelaide.

In Pakistan, the reclamation possibilities of draining and irrigating by tubewells were studied in four very large Salinity Control and Reclamation Projects (SCARPs). From 1959 to 1974, more than 6800 public tubewells were installed in areas of the freshest available groundwater. Shortage of electric power and lack of maintenance caused temporary problems, but the most serious difficulty was that much of the groundwater was not sufficiently free from salts. In SCARP 1, only one-third of the 2000 boreholes were free of salinity or sodicity problems. The others were classed as either marginal or hazardous, and 500 of the latter were later shut down. Although there had been a 2.5-fold increase in agricultural production, the continual recycling of part of the water supply slowly concentrated the salts. In addition, sodium ions from the salts in solution exchanged with calcium ions in the soil to leave impermeable sodium clays. Sodicity can be cured by application of gypsum (calcium sulfate), but this adds to the costs. By 1975, the average water table had fallen to a depth of 2.6 meters, so that waterlogging was effectively cured in the SCARP areas.

In the expanding irrigation areas of India, tubewells play a vital part both in water supply and in drainage. They are particularly valuable in eastern India, where rainfall is sufficient to provide leaching, but not reliable enough for cropping without irrigation. Small private tubewells, some operated by groups of farmers, have proved more reliable in operation than public service wells. Where rainfall is erratic, there is a tendency to excessive pumping and lowering of the water table. There is a clear need for management of the groundwater resource. The management policy, however, should be one of maximum sustainable exploitation of a valuable underground reservoir rather than of arbitrary restriction on the depths of tubewells.

The farmers' role in water management

Although design and construction of drainage canals must be a government responsibility carried out by professional engineers, the productivity of the areas thus drained depends greatly on the organization of groups of villagers to maintain and manage their farm watercourses. This organization occurs in practice on vastly different scales.

Where watercourses are neglected and leaking, the correction and maintenance of earthen channels among farms must clearly be a matter for the farmers themselves. The Mona team in Pakistan succeeded in persuading more than 100 local villages to form water-users associations, which organized their own programs of watercourse reconstruction and maintenance. They needed technical guidance at a simple level from extension staff for watercourse improvement and field leveling. Progress was improved by small subsidies to provide cement for the small concrete structures, but the main action lay entirely within the capacity of the villagers to help themselves. The Mona team found that such organization was successful where all of the village group belonged to the same religious sect. The presence of competing rival sects within a village often led to failures in the field.

In contrast, where the scale of the waterlogging and salinity is such that new distribution systems are needed to match the new drainage canals, a completely different form of organization is needed. In Iraq, the developments described earlier require the provision of alternative housing as a first step. Complete communities are then moved into the new housing and become employees of the construction project for 3 or 4 years. When the channels are completed, the fields are tractor plowed, leveled, and leached by flooding. Three successive crops of salt-tolerant barley are grown. The third crop usually achieves a complete cover. The farmers are then allotted fields in groups of ten to each watercourse; each group elects a headman who organizes maintenance. Seed and fertilizers are made available, and individual farm operations are resumed. The newly resettled areas rapidly become prosperous and unemployed urban laborers are recruited to help with the fieldwork.

Hydrological management in the lowlands

The storage capacity of the sand and gravel aquifers of the lowland alluvial plains can be used in the overall water management of a river basin to balance seasonal shortages of surface water. The groundwater is pumped in dry weather to supplement surface supplies, thereby creating storage space, which is recharged from rainfall and from surplus riverflow. This system has been developed to an advanced level of control in Israel, where the salt concentrations are a vital factor in the management. Fresh water from the snowfields of Mount Hermon, via the Jordan River and the natural reservoir of Lake Tiberius, is pumped through a national conduit of canals, tunnels, and pipelines with a capacity of 20 m^3/sec, to the Negev Desert irrigation developments. Inputs are made en route from brackish groundwater, with the total salinity controlled by dilution.* Tubewell operation in the Indo-Gangetic Plains achieves a similar use of underground storage capacity. Although Britain has relatively ample water supplies, population increase in the Thames valley has so increased demand that the underground storage provided by deep chalk formations is drawn upon in dry summers and is recharged through special boreholes into which river water is pumped in winter.

The importance of storage is emphasized by the response of rice, the most important cereal of the tropics. In spite of the high temperatures of the tropics, cloud cover is heavy in the rain seasons. The yield of rice responds linearly to radiation for some 40 days before harvest.(5) Maximum rice yields and maximum responses to fertilizers therefore occur in the dry season when irrigation depends directly on storage reservoirs.

---

*Brackish water has less than 1 percent of dissolved salts. Saline water has more than 1 percent. Seawater has some 3.5 percent.

Scales of investments needed

A study by the International Food Policy Research Institute, one of the research centers of the CGIAR described earlier in this chapter, analyzed the plans for renovation and expansion of irrigation facilities by the 35 lowest-income countries of the U.N.(10) The countries included India, Pakistan, Indonesia, and the Philippines, together with 23 countries in Africa between the Sahara and the Zambezi River (Table 4). Because India and Pakistan accounted for 78 percent of the total equipped irrigation area and 60 percent of the planned expansion of this group, these two countries are listed separately in Table 5, which summarizes the investment estimates. The advantage of renovating existing systems instead of creating new systems is shown in Table 5.

During the 1980s, India developed and equipped more than 1 million hectares of new irrigation area every year, helped by massive financial inputs from Western countries. The World Bank and the Asian Development Bank are currently investing in some 40 projects throughout South and Southeast Asia.

TABLE 4

Planned irrigation development to 1990 in 35 low-income nations

	Area (million ha)					
		Proposed improvements				
Area	Irrigated 1982	Major	Minor	Proposed new areas	Total invest. 1983 (U.S. $ millions)	Expected annual food[a] increase (000 tons)
India	45	6.7	10.1	14.0	48,247	51,000
Pakistan	14	2.5	3.8	0.4	3,188	5,000
35 nations[b] total	75	11.0	17.8	24.0	71,561	80,000

[a]Wheat equivalents. [b]Includes India and Pakistan.

Source: Oram, P.; Zapata, J.; Alibaruho, G.; and Roy, S. 1979. *Investment and input requirements for accelerating food production in low-income countries by 1990.* Washington, D.C.: International Food Policy Research Institute.

TABLE 5
Irrigation investment needed to produce 1 additional ton of wheat annually ($/ha)

Area	New areas	Renovations Major	Renovations Minor
India	1120	650	440
Pakistan	1840	690	380
35 nations, total	1100	640	420

Source: Oram, P.; Zapata, J.; Alibaruho, G.; and Roy, S. 1979. *Investment and input requirements for accelerating food production in low-income countries by 1990.* Washington, D.C.: International Food Policy Research Institute.

Protection of lowland investments

Land-management issues are so important to the economy and social stability of developing countries, which are heavily dependent on agriculture, that they should be of major national concern. A detailed computer modeling of Pakistan's crops, climate, water supply, and agricultural economy concluded that "adoption of an efficient water allocation and of economic prices could reasonably lead to a basin-wide output increase of 200 percent more than the 1977 level by 1995 if the WAPDA program is carried out. Population increases will limit this to a 34 percent per-capita gain in the income of irrigating farmers."(1)

Management of the resources of land, soils, and water in the lowland reaches of major watersheds involves heavy investment in irrigation and drainage in addition to the investment concentrated in the urban areas. All these investments are at risk if the upper watersheds are so misused that sediment flows damage the lowland installations. A serious component of this damage is the deposition of sediments in the river channels. As the level of the riverbeds is thus raised, floods spread ever more widely, requiring massive embankment works. The Yellow River crosses the densely settled lowland ricelands of China between embankments that have been raised during the course of the centuries. The water level is now 10 meters or more above the land surface. The vigorous and expert soil conservation campaign in the watershed of the Mississippi River came too late to prevent a similar raising of riverbed levels, so that similar high embankments are now a maintenance liability and a flood hazard. (These issues are discussed in the following chapter.)

Control of water-borne disease

Watershed authorities in the tropics have a further and increasingly serious problem in the control and eradication of bilharziasis (schistosomiasis). This is a water-borne parasitic disease for which a few species of aquatic snails are the intermediate hosts to two major species of parasites. This debilitating disease already afflicts 200 million people in tropical areas where irrigation supports dense populations of subsistence cultivators. It is spreading rapidly as irrigated areas are developed in Africa, Asia, and South America. Children are attracted to play in water, and millions thus become infected for life. Drug therapy is effective, after prolonged research by international drug industries, but medical facilities are usually minimal in such areas.

The most important controls are within the field of watershed management:

- Physical methods involve drainage of marshy areas, cleaning of drainage ditches, and straightening of stream banks to eliminate marshy areas.
- Biological methods involve suppressing floating and submerged weeds that favor snails and encouraging predatory fish to suppress snail populations.
- Chemical methods involve adding molluscicides to the water. Copper sulfate is widely used, but oxonyl, an organic compound, is even more effective.

Research is in progress in Zimbabwe(4) and in Nigeria,(12) but the infection of streams and of natural lakes continues to spread the disease. Social discipline in keeping urine and feces away from water systems is a critical element in the control of these diseases.

Disposal of wastes

The disposal of domestic and industrial wastes is usually the direct responsibility of the engineers of the local or state authorities and monitored by the medical or public health staff. These matters are an essential aspect of river management and therefore of watershed management policy.

Sewage treatment for domestic wastes is by storage under anaerobic conditions (that is, without oxygen supply) in deep tanks in which microorganisms break down organic matter into inoffensive compounds. After filtration, the sewage is stored in open lagoons, where oxidation occurs. This is sometimes accelerated by blowing air into the lagoons. Further filtration (and sometimes chemical treatment) produces water clean enough to be returned to the river from which communities downstream are drawing their water supplies. In densely populated watersheds, such as that of London's Thames River, it has long been necessary to reuse water two or three times on its way to the sea.

If untreated sewage is discharged into the river, the oxygen in the water is used up so that fish cannot survive. After remaining in this condition for a century, the Thames has now been protected by treatment plants, and salmon have returned to the river. The cost has been severe. Sewage presents even greater health problems in tropical watersheds where rivers are used for bathing and for laundry.

In arid countries, treated sewage water is sometimes used to irrigate tree plantations or sprayed on lawns in parks. Strong sunlight sterilizes the water within minutes, so that public access need be prevented only during actual spraying.

References

1. Duloy, J.H., and O'Mara, G.T. 1984. *Issues of efficiency and independence in water resource investments.* World Bank Staff Working Paper no. 665. Washington D.C.
2. Easter, K.W. 1975. Field channels: A key to better Indian irrigation. *Water Resources Research* 11:389–395.
3. El-Tobgy, H.A. 1976. *Contemporary Egyptian agriculture.* New York: Ford Foundation.

4. Evans, A.A. 1978. Routine control of bilharzia in irrigation systems. *The Rhodesia Science News* 12:92–95.
5. Evans, L.T., and De Datta, S.K. 1979. The relation between irradiance and grain yield of irrigated rice in the tropics as influenced by cultivar, nitrogen fertilizer and month of planting. *Field Crops Research* 2:1–17.
6. Hopper, W.D. 1976. *Food production in India.* Ottawa: International Development Research Centre. [The Sixth Coromandel Lecture; reprinted by IDRC]
7. Jacobsen, T. 1958. *Salinity and irrigation: Agriculture in antiquity.* Dyala Basin Archeological Report 1957–58. Baghdad: Ministry of Irrigation.
8. Kemper, W.D.; Clyma, W.; and Ashraf, B.M. 1975. Improvement and maintenance of earthen watercourses to reduce waterlogging and increase supplies for crop production. In *Proceedings of the international conference on waterlogging and salinity,* 197–208. Lahore, Pakistan: University of Engineering and Technology.
9. Kumar, P. 1977. *Economics of water management (a study of field channels).* New Delhi: Heritage Publishers.
10. Oram, P.; Zapata, J.; Alibaruho, G.; and Roy, S. 1979. *Investment and input requirements for accelerating food production in low-income countries by 1990.* Research Report no. 10. Washington D.C.: International Food Policy Research Institute.
11. Pereira, H.C. 1985. Irrigation for future world food production: The Gerald Lacey memorial lecture. *ICID (International Commission for Irrigation and Drainage) Bulletin* 34:1–11.
12. Thomas, J.D., and Tait, A.I. 1984. Control of the snail hosts of schistosomiasis by environmental manipulation: A field and laboratory appraisal in the Ibadan area, Nigeria. *Philosophical Transactions of the Royal Society of London B* 305:201–253.
13. Water and Power Development Authority. 1978. *On-farm water management.* Lahore, Pakistan.
14. Water and Power Development Authority. 1979. *Revised action programme for irrigated agriculture.* Lahore, Pakistan: Master Planning and Review Division.

10
Policies of land use for flood abatement

Managing upper watersheds

Watershed vegetation is not enough

Major floods are generated by atmospheric concentrations of water and energy from areas vastly greater than those of watersheds. Good agricultural and forest management can protect soils and moderate stormflow peaks by enhancing temporary water storage. Rainfall is temporarily held on wet foliage as water films and in leaf litter, in soil pores stabilized by fibrous roots, in terrace channels, and in waterways and ponds.

However, in an overwhelming atmospheric event, as when more than an average year's rain falls in 2 or 3 days, all these types of storage rapidly become saturated. Even in heavy forest, water then pours overland in a volume exceeding the capacity of the streams and main river channels, but good land use will minimize the damage. Trees planted along roads, riverbanks, and farm boundaries hold the soil in place. Protection forests on steep slopes shed water but retain soil.

By destroying hillside forests and misusing the cleared areas so that soil stability is degraded, humans contribute directly to flood hazard. Misused slopes are vulnerable to extremes of rainfall; overland flow is accelerated, and soil and debris of rocks and vegetation are torn loose and transported, causing massive deposition and damage in the lower and more developed reaches. The economy of the whole watershed is thus depressed.

Although field techniques for rehabilitation of upper watersheds are known, successful restoration of a damaged watershed only reinstates the former natural measure of flood control. *To ensure freedom from floods for the downstream urban and industrial areas, the capacity for water storage must be better than that afforded by the natural vegetation.* The technologies have been developed. In the upper watersheds, they require engineering construction at substantial, but justifiable cost; in the lowlands, they may require much reorganization of land use, at a far greater cost, and the construction of protective embankments around floodplain areas for emergency storage.

Provision of storage in the upper watersheds

As development in lowland areas proceeds, the growth in investments vulnerable to damage by floods increases the insurance value of upstream protective structures. Check-dams of modest size, fitted with controlled outlets and built high in the watersheds, are an effective solution. Usually the reservoirs remain almost empty, ready to receive and to store for a few days any sudden excess of streamflow. The outflow from the dams can be controlled to avoid the coincidence of peak flows from major tributaries, which is a dangerous source of lowland flooding. Yet the dams are expensive because the sites are usually remote and because the cost cannot be amortized by power generation or irrigation supply.

Some 50 years ago, the U.S. Congress passed flood-control legislation that authorized funding for such flow-regulating structures. Responsibility for their planning and operation was vested in the Soil Conservation Service. The early results were encouraging. In Oklahoma, an 85 percent reduction in flood damage from severe storms was achieved by building 24 regulating dams to control an eroded watershed of only 260 square kilometers. Their capacity was planned to be sufficient to contain the estimated maximum flow probable in 25 years. Larger flows were bypassed by grassed spillways.(7)

Because the probable maximum size of storm, calculated from past records, is only a rough guide, it sometimes is exceeded. An account in 1967 from western U.S.A. reported a similar concentrated pattern of regulating dams to have only halved the flood peak from a severe storm with 200 millimeters of rainfall, which fell on a wet watershed at rates of more than 140 millimeters per hour. The retained floodwater was released for a period of 20 days.(5)

In tropical conditions, the strong seasonal contrasts increase the importance of upstream storage for flood control. Small, hand-built storage structures, such as village "tanks" and farm ponds, serve a similar purpose, but they also supply water for drinking or supplementary irrigation. When sited for erosion control, they may still serve immediate economic purposes.

The program for the rehabilitation of the Siwalik Hills, a heavily populated range of the Himalayan foothills in India, provides a good example. These hills have been denuded of forest; severe overgrazing has been followed by such violent soil erosion that seasonal torrents now spread sand and gravel over fertile lowlands. A series of 50 small earthen check-dams is under construction, with 15 scheduled to be completed in 1987. The dams are less than 10 meters high, with catchments of 4 to 35 hectares. The area commanded for irrigation is about equal to that of the catchments. For such structures, Das and Singh(4) report local benefit-cost ratios of 1.13:1 for 10 years and 1.6:1 for 20 years. The ratios rise to 2.32:1 and 3.14:1, respectively, when downstream benefits are considered.

Flood routing

Except for catastrophic atmospheric events, most storms cover only parts of the watersheds of river systems, causing stormflow on some but not all of the tributaries. If flood-controlling dams on all major tributaries are linked by modern telecommunications, engineers can delay or release flows from tributaries so that they pass down the main channel in succession. Such flood routing is now a critical component of river-basin management, the largest scale on which watershed management policies can operate.

Similarly, many reservoirs planned for power and irrigation have been built with extra capacity for flood storage. In the monsoon climate, the reservoir levels are usually low at the end of the dry season and offer accommodation for flood storage early in the rains. Their subsequent management, to retain some capacity to moderate floods while providing maximum productive storage, calls for highly skilled judgment, which is greatly assisted by modern computer models of the river systems.

Protection of flood-storage reservoirs

In countries where subsistence agriculture has already destabilized slopes high in the watersheds, construction of flood-controlling dams without repair of their catchment areas can give only a brief advantage before the reservoirs are destroyed by sedimentation. How effective are the treatments to control soil erosion in checking the sedimentation of reservoirs? Hydrological evidence is scarce because the poverty and weak public administration associated with such misused areas hamper sustained measurement. The first quantitative evidence came from the U.S.A., which not only made early and large-scale mistakes in land use, but later achieved the awareness and marshalled the resources to correct them. Well-conducted measurements were part of a rehabilitation program in the mountain watersheds of Utah where tree felling and severe overgrazing resulted in disastrous floods and mud slides. Construction of soil conservation structures together with the restoration of grass and tree cover and the control of grazing reduced both the incidence and severity of summer floods and soil transport.(1) In time, the stability was found to increase.(2)

The classical watershed development program of the Tennessee Valley Authority demonstrated both restoration of soil and water management and the economic recovery of an eroded and poverty-stricken landscape. The investment in storage structures, mainly for power generation, is impressive (Fig. 37). Their protection was essential. Critical studies were made for several years on eroded subwatersheds scheduled for restoration in order to measure the changes achieved by watershed treatments. The results emphasized the effectiveness of the cutoff drains and the contour cultivation techniques of the U.S. Soil Conservation Service and of the reforestation of steep slopes. The studies also showed that there is a price to be paid in terms of water to fill the reservoirs in dry weather. Stormflow peaks were reduced by 90 percent and sediment transport was reduced by 96 percent, but the total yield of water was halved.(14, 15) Watershed research and soil conservation studies in East Africa provided evidence

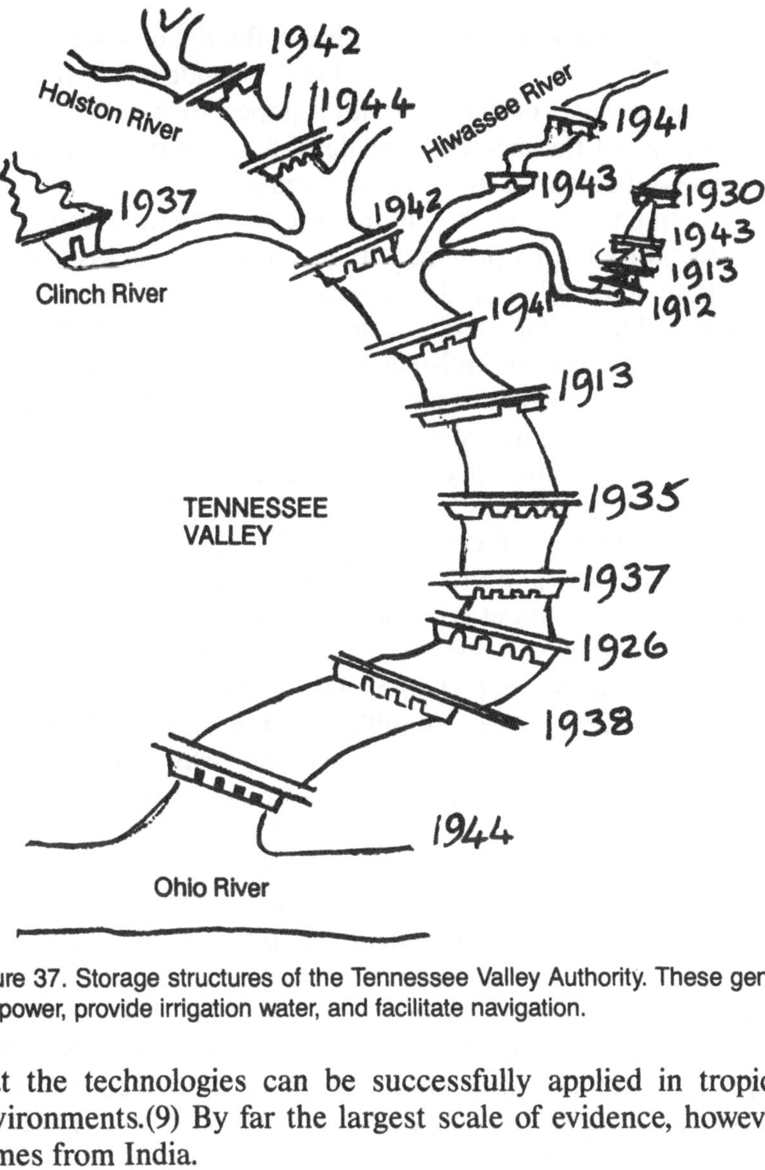

Figure 37. Storage structures of the Tennessee Valley Authority. These generate power, provide irrigation water, and facilitate navigation.

that the technologies can be successfully applied in tropical environments.(9) By far the largest scale of evidence, however, comes from India.

Case studies from the Himalayas and their floodplains

Flood control works in the upper watersheds

The grandest natural theater in which all these problems are being played out is the Himalayan range. Steep, high mountain watersheds supply the crowded alluvial floodplains of the Indus and the Ganges river systems. There is evidence here of the increasing misuse of the mountain slopes by rapidly growing populations; of the increasing scale of flooding on the plains; of the surveyed extent of the erosion damage; and of the success of land treatments. Figure 38 demonstrates how some 40 million hill dwellers are sited to provide either stability or devastation for more than 400 million on the plains.

Destruction of forest in the upper watersheds of India and Nepal, followed by land misuse, has brought into sharp focus the hazards to downstream users. In India, the government response has been to provide increased funds for soil conservation. During the first five of India's 5-year plans, a total of U.S.$600 million was provided, but in the Sixth Plan (1980–1985) some $530 million was budgeted, of which about $410 million was from the state governments and $120 million from central government.

As part of the flood-control policy, India has in the past 2 decades established major river-valley projects for watershed improvement to protect 31 important reservoirs on flood-prone rivers. Their total area is 79 million hectares.(3) The Ministry of Agriculture's Soil and Water Conservation Division in 1980 reported a survey on 21 of these reservoirs for which adequate sediment sampling records were available. The survey revealed 45.8 million hectares of eroded land in the catchment areas of the 21 reservoirs to be in need of rehabilitation, of which less than 2 million hectares had been treated. Comparison of the sedimentation rates before and after dam construction showed that in 17 reservoirs the rates had more than doubled and that four of these reservoirs were filling up at 6 to 20 times the initial rates.(8) In only one catchment area had the sedimentation been reduced to below the designed level. This was the Machkund Reservoir, whose 200,000 hectare watershed had been completely treated.(10)

The critical point is that where the treatments had been completed, the sedimentation problem had been overcome. Table 6 demonstrates that where only part of the critical area of erosion had been treated, partial reduction in the rates of sedimentation was achieved

Further evidence of the reduction of both peak waterflows and

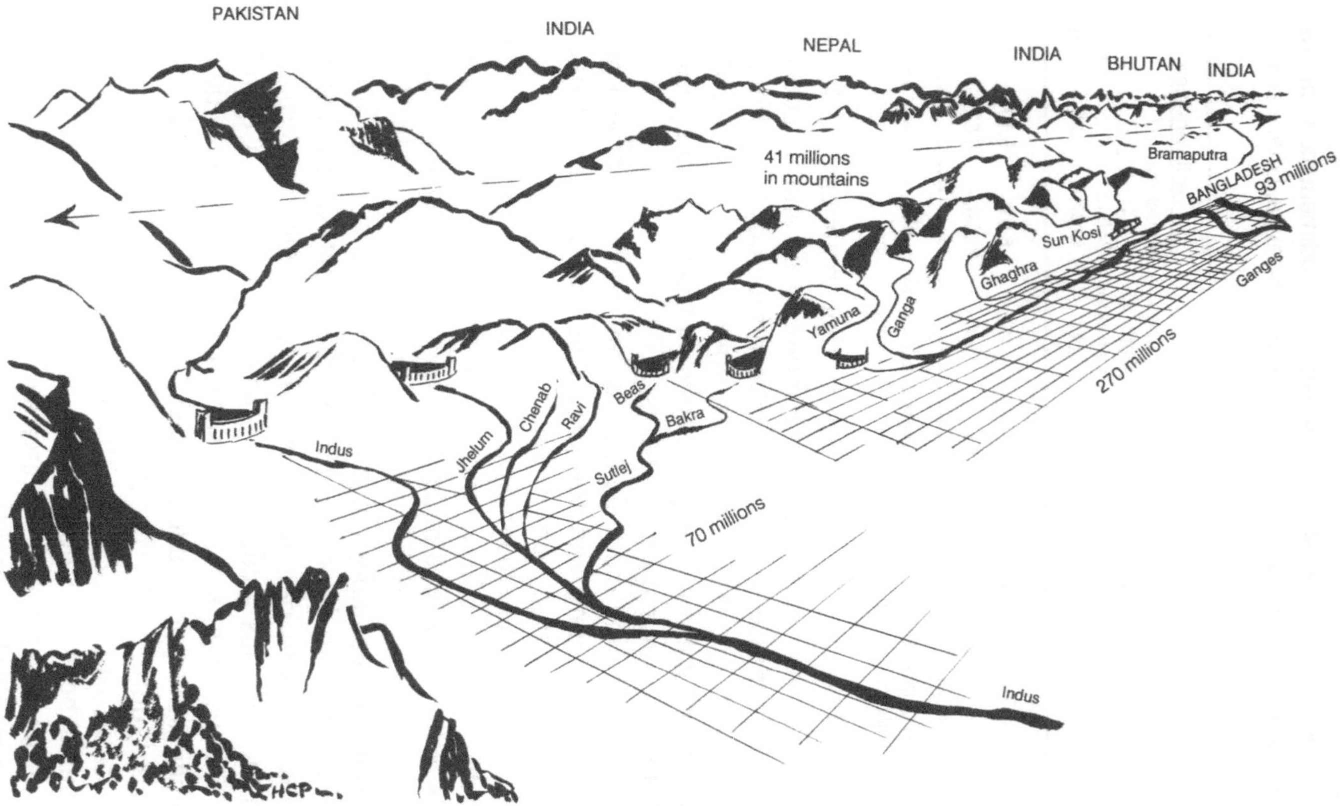

Figure 38. Irrigation of the Indo-Gangetic Plains supports more than 400 million people. They are at risk from the misuses of the mountain watersheds by about 41 million people who need help in restoring their spectacular but vulnerable environment.

TABLE 6

Effects of soil conservation treatment on sedimentation in two reservoirs (Bhakra, Himaschal Pradesh; and Ramganga, Uttar Pradesh) in India

					Sedimentation		
					Annual (ha.-m/ 100 sq km)		
Reservoir	Year constructed	Catchment area (000 ha)	Area of critical erosion (000 ha)	Portion of critical area treated (%)	Before treat.	After treat.	Reduction (%)
Bhakra	1959	1820	537	23	6.91	5.19	25
Ramganga	1974	315	104	40	22.00	16.80	24

Source: G. P. Gupta. 1980. Soil and water conservation of river-valley projects. *Indian Journal of Soil Conservation* 8:1-7.

sedimentation rates by soil conservation treatments of Indian watersheds was summarized by Subramaniyan and Samuel.(11) As in the Tennessee valley, the treatments included the afforestation of steep slopes, cutoff drains, grassed waterways, contour plowing, and gully stopping by sediment traps. It is not possible to distinguish their separate effects.

Such restoration methods increase the productivity and profitability of farming and forestry. This was illustrated quantitatively by a study on 16 small watersheds reclaimed from ravine lands in the eroded Siwalik Hills near Dehra Dun, India. Forested slopes had been stripped and overgrazed to bare wastelands. With about 1800 millimeters of annual rainfall, the ravine lands were a source of floodflow and erosion debris. In the 1950s, experimental replantings of gullies having shallow rocky soils on slopes from 20 to 45 degrees (36 to 100 percent) were made with hardwoods *Dalbergia sissoo* and *Acacia catechu*, interplanted with the vigorous fodder grass *Chrysopogon fulvus*. Control of soil and water was restored, with a 90 percent reduction in sediment transport. Building poles, fuelwood, and fodder were assessed, during a 17-year period, to have produced an overall benefit-to-cost ratio of 2.7:1. To this must be added a substantial economic benefit to downstream users when a large enough proportion of the catchment area had been treated.(13)

Thus many of the essential techniques to prevent the destruction of reservoirs by sedimentation have been tested and established in India, but the problem is growing. Correction depends on convincing many millions of subsistence farmers of the advantages of changing traditional practices. Training of the necessary technical staff to sustain a national drive for soil conservation by better agriculture is itself a major undertaking. It has not yet reached a scale adequate to meet even the current needs of existing projects.(12)

Sediment over the spillways

In the majority of the 21 reservoirs whose sedimentation rates have been studied in India, misuse of the main areas of their watersheds has persisted and flood peaks and sedimentation have thus continued to increase. The consequences have not been limited to the destruction of reservoir storage capacity. All dam walls are protected by spillways, which will pass high flows around the dam rather than over the crest when the reservoir is full. Thus, when floodflows are fast and turbulent, much of the sediment and debris are carried over the spillway, causing damage downstream.

A large-scale example is provided by the Khosi River in the Himalayas. The river drains 61,000 square kilometers of mountains whose easternmost 200 kilometers are within Nepal's boundaries. Several large tributaries combine to escape through the single exit of the Khosi gorge and violently discharge onto the Terai (the Nepalese plains). Characteristically, these major torrents are unstable in flat country, and during the centuries, there have been many changes, with heavy damage. The flows have now been constrained by embankments and checked by the Hanumanaga barrage at the border with India. Built in 1959, this major storage and regulation structure offered important irrigation opportunities in Bihar state, India. Large investments were made in a new canal network in Bihar, but the land misuse in the upper watersheds in Nepal was not corrected, and the watersheds were not treated for erosion control. By 1977, some 20 percent of the new canals had been blocked with sediment.(16) In spite of heavy expenditure for cleaning the canal systems, the problem persists.

Flood-storage areas in the lowlands

Throughout geological history, rivers in the flatter lowland reaches have overflowed their banks and deposited sediment. When watersheds were uninhabited and protected by vegetation, the sediment came from the undercutting of the banks and scouring of steep riverbeds, together with local disturbances such as avalanches and landslides. The floodplains thus evolved as natural storage areas for floodflow and reception areas for sediments.

These alluvial plains, when drained and protected from floods by earthen embankments, usually have good agricultural potential. They also offer low-cost routes for road and rail, and sites for urban development. Settlements protected by embankments date from more than 5000 years ago. Yet, by denying flood storage to the river system, such embankments increase the height of the flood. Because the rate of flow diminishes in the flatter lowland reaches sediment deposition increases. If sediment can no longer be deposited on the plains, much more is dropped on the riverbed, thus raising the level, with the result that banks have to be raised or they will be overtopped. The long-term consequence is that the river runs in an elevated bed above the flatlands, as do the Yellow River and the Mississippi River (Chapter 9). As the Gangetic floodplains become increasingly populated and developed, the need to avoid such consequences has deeply concerned the central government of India.

Increasing flood hazard on the Gangetic Plains

In 1978, the largest flood ever recorded on the Yamuna swept through part of New Delhi. Fortunately, a well-developed flood warning system gave the capital city 3 days notice, and no deaths were recorded. A high-level government working group investigated the event and in the "Mukerjii Report," issued in 1978, found that evidence for a very serious and continual increase in flooding was unequivocal: From 1953 to 1971, the area affected by floods was 25 million hectares; from 1971 to 1977, the area was 28 million hectares; in 1978 the area was 34 million hectares. Meteorological records showed that this change was not due to any progressive increase in rainfall.

Protective construction in the floodplains

The historical response to periodic flooding has been to protect the threatened plains by earthen embankments. By 1947, India had protected some 3 million hectares with more than 3000 kilometers of embankments. Severe floods in 1954 resulted in the National Flood Control Programme, which led to the building of more embankments as a short-term policy and more flood-storage reservoirs as a long-term policy. By 1978, a further 9.8 million hectares were protected by 10,780 kilometers of new embankments and 19,000 kilometers of new drainage channels. The protected area enclosed 5000 towns and villages.

For 20 years, the central government has been urging the state governments to set up hydrographic networks to quantify and improve the routing of seasonal high flows across the plains and to establish floodplain zoning and control of development. In 1976, the Ministry of Agriculture and Irrigation appointed the National Commission on Floods to conduct an in-depth scientific study of flood control, including floodplain management for the whole of India. This monumental task is still in progress.

Integration of watershed measures for flood abatement

Flood-control measures have thus begun, historically, at the wrong end of the rivers. Throughout the developing countries, the more wealthy communities of the lowlands have invested in protection structures while the rural areas have been left to traditional methods of livelihood. The evidence is now clear that population increases in the upper watersheds have invalidated these traditions of land use: Soil and water problems from land misuse now affect the complete watershed community. Flood damage will be effectively reduced only if people recognize that the watershed functions naturally as a single system.

When the soil protection and flow retardation provided by the forest have been eliminated by agricultural settlement, flood-control storage structures will often be needed as insurance against flood damage to the lowlands, even if soil conservation discipline has been established. Floods have been moderated in

Iraq by diverting floodflows into natural depressions. Barrages across the Euphrates at Ramadi and across the Tigris at Samarra send floodwater along canals to be stored in Lakes Habbaniya and Tharthar, respectively. Where floodplain storage has been reduced by river embankments, emergency flood-relief areas may be necessary to impound flood waters temporarily. These areas can be used as pastures that can be rapidly cleared of livestock before controlling gates are opened. This method is long established for control of high water levels on the lower Mississippi. However, on the Gangetic Plains, where development has not been guided by overall land-use planning for flood control, the problems of clearing land for temporary flood impoundment areas are formidable, and the task of the National Commission is inevitably slow and difficult.

A practical example of the overall planning for flood control and irrigation development in a semiarid developing country is the Wadi Najran project in southwestern Saudi Arabia. After regional surveys, intensive field measurements were undertaken in order to construct computer simulations of rainfall, evaporation, runoff, infiltration, and the distribution and quality of groundwater. The studies showed that floods could be controlled by a dam built upstream of the alluvial plain. This would increase infiltration and thus recharge groundwater storage. An increase of 35 to 40 million cubic meters per year in the yield of groundwater would permit development of irrigated agriculture from a series of boreholes sited to draw on the recharged storage.(6)

References

1. Bailey, R.W.; Craddock, G.W.; and Croft, A.R. 1947. *Watershed management for summer flood control in Utah.* Misc. Pub. 639. Washington D.C.: U.S. Department of Agriculture.
2. Croft, A.R., and Bailey, R.W. 1964. *Mountain water.* Ogden, Utah: U.S. Forest Service.
3. Das, D.C.; Bali, Y.P.; and Kaul, R.N. 1981. Soil conservation in multi-purpose river valley catchments. Problems, programme approach and effectiveness. *Indian Journal of Soil Conservation* 9:5–26.
4. Das, D.C., and Singh, Shamsher. 1980. Small storage works for

erosion control and catchment improvement: mini case studies. *Proceedings of the international conference: Conservation 1980*. Bedford, U.K.: Silsoe College.

5. Hartman, A.M.; Ree, W.O.; Schoof, R.R.; and Blanchard, B.J. 1967. Hydrological influences of a flood control program. *Proceedings of the American Society of Civil Engineering, Hydraulics Division* 93:17–25.
6. Jones, K.R. 1985. *The Wadi Najran Project, Saudi Arabia*. Wallingford, U.K.: British Hydrological Society.
7. Kautz, H.M. 1955. The story of Sandstone Creek watershed. In *Yearbook of agriculture 1955*, 210–218. Washington D.C.: U.S. Department of Agriculture
8. Ministry of Agriculture. 1980. *Statistics on soil conservation in India*. New Delhi: Soil and Water Conservation Division.
9. Pereira, H.C. 1973. *Land use and water resources*. Cambridge: Cambridge University Press.
10. Samuel, J.C.; Mukherjee, B.K.; and Das, D.C. 1980. Reduction in sediment production due to soil conservation measures in Machkund catchment. In *First national symposium on soil conservation and water management*. Dehra Dun: Indian Association of Soil and Water Conservation.
11. Subramaniyan, S., and Samuel, J.C. 1982. Hydrologic and sediment response variations in watersheds. In vol. 1 *International symposium on mountainous watersheds*. Roorkee, U.P., India: University of Roorkee.
12. Tejwani, K.G. 1984. Reservoir sedimentation in India. *Water International* 9:15–154.
13. Tejwani, K.G.; Gupta, S.K.; and Mathur, H.N. 1975. *Soil and water conservation research (1956–1970)*. Publication no. 68. New Delhi: Indian Council of Agricultural Research.
14. Tennessee Valley Authority. 1962. *Reforestation and erosion control influences upon the hydrology of the Pine-Tree Branch watershed (1941–1960)*. Knoxville, Tenn.
15. Tennessee Valley Authority. 1963. *Parker branch research watershed project report 1953–1962*. Knoxville, Tenn.
16. Verghese, B.G. 1977. *Gift of the greater Ganges*. New Delhi: Coromandel Fertilizers.

11
Responsibilities of governments and aid agencies in watershed rehabilitation

National policy decisions

The preceding chapters have explained the urgency of restoring soil and water stability in the watersheds of the tropical and subtropical developing countries. The major policy decision is that rural communities must be instructed and helped to grow fuelwood and forage. This basic decision will give opportunities to restore steep land to hydrological stability and to return misused wastelands to vegetative production, thus combining watershed improvement with better living standards for the rural poor.

To bring about these changes, urban centers will have to increase support for remote rural areas by providing communications and housing for resident technical and administrative workers. Protection must be given from invasion by migrants and from raids by illicit urban collectors of fuel and timber. Production of fuelwood and forage for supply to the urban areas must replace the competitive destruction of tropical forests. These policy decisions and executive actions can be taken only by national governments as political leaders become aware that seemingly remote rural problems are profoundly important to the national economy.

The number of international meetings held, papers published, resolutions issued, and declarations of environmental policy signed would suggest that developing countries have indeed made a positive response to their watershed problems. Yet all of the technical assessments indicate a rapid worsening of the situation. The contrast is painfully clear to travelers in the high watersheds. Many of the worst-affected developing countries have environmental authorities, interdepartmental coordinating councils, and special agencies, all with offices and staffs in the cities. The small number of technical and administrative staff members housed and working in the remote rural areas where the action is required has received far too little reinforcement.

The role of international aid

External aid for the improvement of tropical watersheds has been overemphasized. With the support of aid agencies, constructive land rehabilitation and community organization projects are in progress in several hundred valleys throughout the tropical world. Many have made good progress, but their impact is limited by their pepper-pot pattern. The improvement of a scattering of small subwatersheds can have little overall hydrological effect while land misuse on a large scale continues all around them. They are essential to give both watershed communities and government authorities confidence that the technologies can succeed, but without strong government support of national decisions on land-use policy, the subwatershed improvements will be transient.

Watershed rehabilitation in Nepal, carried out in rural development projects within subwatershed boundaries, provides clear examples of the issues. A national allocation of major watersheds to each of several bilateral and international donors has been made (Fig. 39). Within each, a subwatershed is being developed. Nine aid agency field teams are served by a small technical team of WB/UNDP/FAO silviculturalists and sociologists in community forestry development.

The technologies are being effectively demonstrated. Enough is known from previous projects in agriculture and forestry and from effective agricultural research done at the former British Gurkha resettlement centers at Lumle and Pakribas. The social patterns are, however, so complex that improvements must be made through trial and error. A solution in one valley may fail in

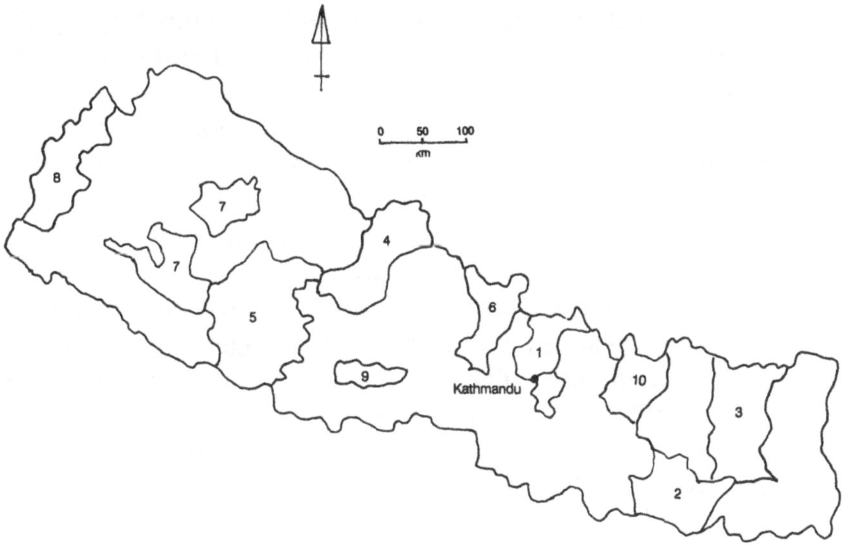

Figure 39. Locations of rural development projects in Nepal served by the World Bank/UNDP/FAO Community Forestry Development team.

Projects	*Donor agencies*	*Cost (U.S. $ millions)*	*Begun*
1. Rasuwa/Nuwacot	World Bank	11.9	1976
2. Sagamatha/Udayapur	ADB	30.2	1978
3. Khosi	UK	9.4	1979
4. Myagdi/Mustang 5. Rapti 6. Ghorka	USAID	33.7	1980
7. Karnali/Bheri	CIDA	12.5	1980
8. Makahali Hills	World Bank	13.0	1980
9. Tinau Khola	Switzerland	10.0	1980
10. Lamusangu/Jiri	West Germany SATA	6.2	1980

Source: B.D. Strestha, P. Van Ginnekan, and K.M. Sthapit, *Watershed condition of the districts of Nepal* (Kathmandu: DSCWM, Ministry of Forests).

the next. Constraints may be tribal traditions, or religious groupings, or landownership and tenure in which government departments can be as difficult as private owners.

The criterion for success is not primarily the progress within the treated area of each project—that outcome should be assured by the concentration of technical advice, leadership, and financial support. Rather, success will be judged ultimately by the extent to which improved patterns of land use are taken up in the surrounding major watershed.

There are two important dangers in this approach, however. The first is that a high proportion of the host country's technical field staff is concentrated in the project areas as counterparts to the assistance agency specialists. The tasks are essentially long term, but the patterns of international funding are characteristically short term, 3 to 5 years. Some watershed projects are planned for 15 years in 5-year renewable stages, but continuity depends more on the availability of funds than on the progress of the projects. When funding of a project ceases, the abrupt departure of government technicians to serve as counterparts in new projects elsewhere has led to the loss of much expensively won progress. Attempts to remedy the shortages of trained fieldworkers by substantial allocations of funds for training overseas have been largely nullified in watershed work by the small proportion of the returned trainees to be found stationed in the field.

The second danger lies in the impression that improvement of the land use in hill or mountain watersheds must depend on external aid. Even if the whole of the World Bank's annual lending, of the order of U.S. $9 billion a year, was allocated to watershed restoration in the Third World, only a small proportion of the task could be undertaken. The job can be accomplished, however, by national governments willing to tackle their own problems.

The role of governments

Self-help by rural communities

Rural communities have shown they are able to change the patterns of their husbandry of crops, animals, and trees without massive outside assistance. Small payments made directly to rural farmers or landless laborers for planting fodder and fuelwood have been shown to be effective in India (Chapter 6), Nepal (Chapter 7), and Indonesia (Chapter 8). Simple reorganizations of pasture management were effective in Uganda and in Kenya (Chapter 7). If field workers of several government departments are involved, the costs will be higher, but they are essentially fulfilling the departmental responsibilities and should not need external support. Changes in farming practices in Java required only brief and modest support for 1 or 2 years while farmers adopted contour cultivation and fuelwood planting, with help from several government departments (Chapter 8).

An important element is the strengthening of village community organizations for self-help. Some of the watershed projects in Nepal have initiated village organizations of motivators as nontechnical auxiliary extension staff with only a few weeks of training.(4) Such supplementary organizations may make it possible for technical advice actually to reach the several hundred villagers for which each government extension worker is responsible. Expansion of the government staff on the required scale would be prohibitively expensive and not necessarily more effective. Hundreds of nongovernmental organizations are active in extension work and are winning increasing support from donor agencies. Many have enthusiasm but as yet rather little practical experience.

Self-help demands correct technology. External aid projects are committed to demonstrating and propagating well-tried methods, which are often new to the rural community. The demonstrations involve local verification of technologies; verification is essential because well-known crop varieties and planting methods may fail when tried in a new locality. The destruction of tree seedlings by termites on one soil type and the absence of attack on another soil are the kinds of technical issue that determine success or failure. It is therefore critical that nongovernmental organizations and teams of auxiliary community workers both receive close supervision from the technical departments. For farmers at subsistence standards of living, technical mistakes are not readily forgiven.

Technical help in rural areas

More continual and readily available technical help in remote rural areas is a key to progress in management of upper watersheds. Such topics as soil conservation, forestry production of fuel and forage, intensification of cropping where topography permits, and the management of stall-fed livestock require on-the-spot technical advice and competent public administration. Supervision of government credit for small farmers to purchase improved seed, fertilizers, and other inputs and the organization of input supply and distribution require an educated cadre resident in remote areas. This will be most readily achieved if the best local students are given the necessary education because government workers and their families from the more urbanized areas of the lowlands are rarely content to live in the hills.

The training-and-visit system has been developed in India with World Bank support(1) in order to improve the deployment of subject-matter specialists into rural areas. Funds for housing and for transport have been critical inputs to get such teams of "muddy-booted field advisers" beyond the areas easily accessible from towns. Such field instructors at graduate level also play a critical role in modern Western agriculture. In addition to housing, other necessities are a safe local water supply and availability of food and other domestic supplies. Markets in those districts most in need of help may have little food for sale during the lean seasons. In the Far West region of Nepal, I have seen rice flown in by light aircraft to sustain the technical and administrative cadre during periods of shortages.

Substantial salary increments and career credit for service in remote rural areas are basic necessities that developing countries have been slow to acknowledge. Costs of living are, in practice, high in areas of extreme transport difficulties, but all too often cost-of-living allowances are larger in the cities (where the rules are made), so that urban-based staff enjoy privileges and amenities. Correction of these serious anomalies is an urgent aspect of watershed management.

Technical help with alternative sources of energy

Simple techniques for use of solar energy are especially promising for reducing demand for fuelwood in semiarid watersheds. In Gansu province in northwest China, I have seen very poor peasant farmers using simple, portable solar cookers designed by the local university, with funding by the United Nations Development Programme. Thin dishes of concrete, about 2 square meters in area and 50 millimeters thick, are cast on accurate molds. The concave surfaces are coated with bitumen into which are pressed small squares of mirror glass. This reflecting surface focuses sunshine onto a suspended kettle. Two liters of water reach the boiling point in 20 to 30 minutes. The cost is U.S. $70, but this is reduced by government subsidy to about $10. Although there have been problems with deterioration of the mirror coating, the cooker is much in demand.

In South and Southeast Asia, experimental development of water pumps powered directly from batteries of solar cells on a frame about 2×2 meters is in progress, funded by the United Nations Development Programme, but costs are still high and management requirements are not yet simple enough. Solar energy can be of direct help to the watershed manager in drying products to reduce transport costs. Many tropical examples are available, such as producing copra from coconuts or *biltong* (sun-dried meat) from game harvesting, but other techniques could be introduced by simple technology transfer.

For example, in the Indian Himalayas, an attempt to reduce the rate of watershed forest destruction in Himachal Pradesh involved encouraging farmers to plant apple trees. The result was horticultural success but ecological damage. The apples yielded so well that three new sawmills were set up in the pinewoods to produce the thousands of apple boxes needed to transport the harvest. Production from 70,000 hectares of orchards requires 150,000 cubic meters of standing timber every year. Thus 1 kilogram of wood accompanies every 2 kilograms of fruit down to the plains. One technical solution was the introduction of paper cartons, which reduced the box weight by 80 percent, but the cost was too great.

A more energy-efficient solution would be to remove the water from the fruit. A simple technique, long used in southern Europe and the Middle East, is to sun-dry apples. It requires manual peeling, coring, and slicing of the fruit, immersing it immediately for 10 minutes in a 1 percent solution of common salt, and then threading the apple rings on a stick to dry in the sun.

Perhaps the most elegant way of reducing the energy costs of transport that I have encountered was in the middle mountains of Nepal. Here wild peach trees had been successfully top-grafted with edible varieties and yielded well, but a 3-day porterage or pack-saddle transport over steep trails prevented marketing. A training visit by the local agricultural officer to France, however, led to the construction of a huge but simple still by Nepali coppersmiths to produce peach brandy. Marketing of the fiery liquor has presented no problems. Customers fly to the nearest airstrip and walk up the mountain trail to buy the product.

Thousands of installations that produce methane by fermentation of manure in closed tanks are now operating in tropical countries. Such biogas equipment performs well under competent care. It is best suited to village operation where some basic mechanical knowledge is available. The difficulty of continually collecting manure and of controlling the distribution of the gas has sometimes led to abandonment of equipment, but as the technology passes into rural general knowledge, it will certainly spread. The generation of biogas not only replaces fuelwood; it prevents the burning of the manure while conserving nutrients in the residues from fermentation, which are applied to cropland.

The improvement of charcoal kilns (Chapter 5) is of direct interest to watershed managers because it reduces by two-thirds the number of trees felled to produce 1 ton of charcoal. Such simple equipment as biogas fermenters and charcoal kilns can be

Road building in steep, remote mountains is prohibitively expensive. Market products have to be adapted to the transport available.

made locally in most developing countries. The need is local technical leadership in the rural areas rather than overseas aid.

Watershed managers should encourage water engineers to help in the improvement of indigenous waterpower devices. Traditional small wooden waterwheels in the Himalayas simultaneously grind maize and turn prayer wheels. In Nepal, an initiative by UNICEF, funded by the Asian Development Bank, has produced the "microhydel" multipurpose power unit. The unit consists of a steel Pelton wheel housed in a portable steel box; a second box mounts a grinding stone and a power takeoff to operate a small dynamo that produces sufficient electricity for lights (Fig. 40). An 80-liter-per-second flow at 3 meters head yields 2.4 horsepower at the pulley. The complete set, made in Kathmandu, costs U.S. $1,200 and is selling rapidly.(3)

Minihydel is the term for a larger water-power unit for locations having a streamflow of about 250 liters per second. Under a 5-meter head, a dynamo run by this unit will produce about 10 kilowatts for local electric lighting, milling, or oilseed crushing. Both these types of small hydroelectric equipment can be dismantled into components that porters can carry up mountain trails. The installation of such equipment can help watershed managers because it arouses interest in the care of the watercourse. As standards of living rise, the use of small hydraulic-ram pumps, which will pump water up steep banks from fast-flowing stream channels, should be encouraged, especially for supplying water to stall-fed cattle.

Sustained research and development in laboratories and experiment stations is necessary to provide a continual flow of improvements to these simple field technologies as better materials and methods become available. Support for research comes increasingly from direct contracts by aid agencies for problem-solving work on well-defined practical objectives. Nevertheless, the major task remains for governments to encourage watershed communities to apply knowledge and techniques that have long been in practical use.

Technical help with modification of population growth

The satisfaction with which the staff of a watershed improvement project can regard an orderly valley with wooded steep

Figure 40. Minor water power high in a watershed. Minihydel (left) has a turbine and generator. It normally needs road access, but the equipment can be dismantled and hand carried. Microhydel (right) has a Pelton wheel and power takeoff for a grinding mill, oilseed crusher, or a 12V dynamo. This equipment can be carried by porters.

lands, terraced slopes, and irrigated riverside flats, won back from an eroded and poverty-stricken area, is tempered with the gloomy realization that another 10 or 15 years of present rates of population growth will cancel these hard-won advantages. The need to give family planning campaigns genuinely high priority

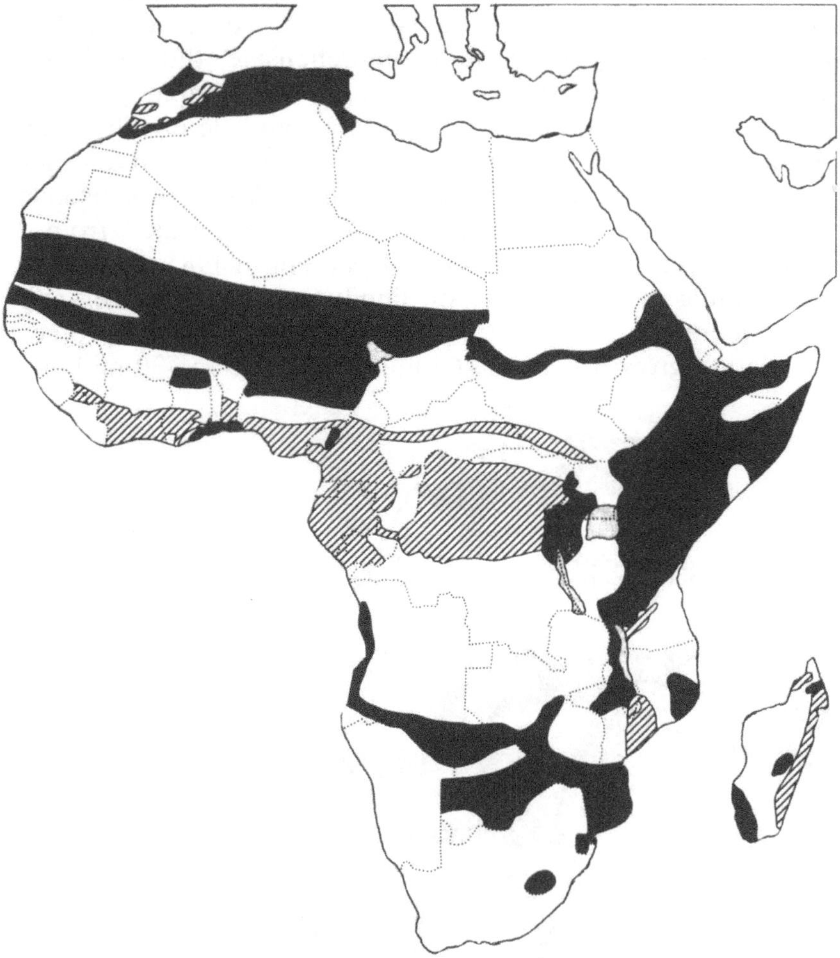

Figure 41. Africa has the land capacity to support people at low levels of input of fertilizers and pesticides. Shaded areas have high capacity (100 persons per square kilometer). Black areas are where population already exceeds the carrying capacity at present levels of inputs. *Source:* FAO, *Population: Potential supporting capabilities of lands in the developing world*, FAO Technical Report INT/75/13 (Rome: 1984).

and more than token resources is the most urgent watershed issue for national governments and donors. In Africa, large sectors of the land area cannot adequately support even today's population (Fig. 41) with the present low-input agriculture. By the end of the century, the growth rates that are implicit in today's age structures will have produced 200 million people above carrying capacity. The World Bank devoted the *World Development Report 1984* to a comprehensive study of the difficulties imposed on developing countries by rapid population growth.(5) These difficulties are especially acute in sub-Saharan Africa, where food production per capita has declined during the past 15 years in all countries except South Africa. A subsequent study(6) reports that population growth accelerated from 2.8 percent a year in 1970–1980 to 3.1 percent in 1985. Assuming a quadrupling of expenditure on family planning services and a consequent decline in fertility 5 to 10 years later, the projected population growth is from 460 million in 1985 to 730 million by

Figure 42. The scale of the African areas of food deficit can best be appreciated by comparing Africa with the size of other regions.

2000. Not even a rate of resource growth as high as that achieved by the developed world during the past 50 years would be enough to improve per capita incomes under such circumstances.

A thorough investigation by FAO with the help of the International Institute for Applied Systems Analysis has estimated the potential for food production for each African country.(2) Although the indications are that the continent has a theoretical capacity to grow food for three or four times the present population, the current trends are ominous. Should prevailing national policies continue to be biased against agriculture, with neglect of provision for inputs and for incentives to farmers, then the divergence of population growth from food production "will be unsustainable ecologically, economically and politically. The food situation in Africa by the year 2010 would be worse than at the height of the 1983/85 famine."(2) The scale of the African areas of food deficit can best be appreciated by the comparisons in Figure 42. The World Bank study(5) concluded that "because poverty and rapid population growth reinforce each other, donors and developing countries must cooperate in an effort to slow population growth as a major part of the effort to achieve development."

References

1. Benor, D., and Harrison, J.Q. 1979. *Agricultural extension: The training and visit system*. Washington D.C.: World Bank
2. Food and Agriculture Organization. 1986. *African agriculture: The next 25 years*. Rome.
3. Nakarmi, A.M., and Bachman, A. 1983. *Multipurpose power unit with horizontal water turbine*. Kathmandu: UNICEF.
4. Swiss Association for Technical Assistance. 1983. *Integrated hill development project, Nepal*. Kathmandu.
5. World Bank. 1984. *World development report 1984*. Washington, D.C.
6. World Bank. 1986. *Population growth and policies in sub-Saharan Africa*. Washington, D.C.

12
Sources of information and recommended reading

For full-time professional research workers, the continual study of the literature must be a way of life. Whatever problems they are working on are also being studied by many others. To fall out of current awareness can only lead to much waste of time and of research facilities. The following notes, however, are written not for the researcher but for those concerned with watershed management. Watershed managers certainly should keep abreast of progress elsewhere in forestry, soil conservation, hydrology, and the adaptations of technology illustrated in the previous chapter. The soil and water aspects of their problems are highly technical fields and do not yield readily to amateur solutions. Engineers, agronomists, foresters, and livestock and veterinary specialists need to be called on for advice and help. Having had that advice, those at the more remote stations will often enough have to put it into practice themselves.

Descriptions

Environmental issues have been generating an ever-growing volume of writing for the past 20 years. Many papers in this area are aimed at political targets; indeed there are now "Green" political parties in Europe. Many others discuss global issues and are aimed at fund administrators or government policymakers. Watershed managers in developing tropical countries may search these papers in vain for the type of information they need. The test of usefulness is whether the sources of data are cited (and if so whether these come from factual reports or from other general essays) and whether the tree species, rainfall, altitude, and so forth are given or are likely to be found in the references.

Scientific abstracts

Whereas the technical adviser must keep up with developments in such areas as veterinary science or in water-pump design, for which the research literature is fairly well organized, the watershed manager has to scan a wide range of subject matter. For both, however, the scientific abstract services are the starting point. Journals have become so costly and have proliferated so widely that even large libraries have difficulties in dealing with the numbers. Several new scientific journals appear every month. Publication in a refereed journal is still the only effective way of transmitting research results.

Most journals require authors to condense their findings into a quarter-page summary. Abstracting services compile abstracts from a broad range of journals and after checking, expansion, or pruning, as necessary, they are published in a highly organized form. For watershed management in tropical developing countries, the Commonwealth Agricultural Bureaux (Wallingford, Oxon OX10 8DE, U.K.) provide the most useful service in agriculture and forestry. *Forestry Abstracts, Soils and Fertilizers, Field Crop Abstracts, Herbage Abstracts,* and *Agricultural Engineering Abstracts* are the most relevant. For irrigation there is excellent coverage by the International Irrigation Information Center (Volcani Center, PO Box 49, Bet Dagan, Israel).

Abstracts are intended to guide readers to papers. They are not substitutes for the papers themselves. The authors may express exceptions, cautions, and limitations about the results that are not retained in the abstract. Buying photostat copies from overseas sources is slow and expensive. Microfiche copies, in which 48 or 96 pages of a journal are reproduced on a postcard-sized transparency, are practical and can be airmailed economically. All up-country stations in developing countries should be equipped with microfiche readers.

For progress in rural water supplies and equipment the

monthly magazine *World Water* is informative and easy to read. It is produced by the Institution of Civil Engineers (ICE) (Telford House, Old Street, London EC1 U.K.). There are also many informative articles on major engineering developments in the tropics in *The New Civil Engineer,* which is also published by ICE.

Journals

For the manager who is not active in research, the journals present fragmented scientific topics; few papers are concerned with land use on whole watersheds. Useful and interesting papers on watershed management are occasionally found in the hydrological journals, such as *Journal of Hydrology* (Elsevier, Amsterdam), *Hydrological Sciences Journal* (Institute of Hydrology, Wallingford, U.K.), and *Water Resources Research* (American Water Resources Association, Urbana, Illinois, U.S.A).

Agricultural techniques for tropical watersheds can be found in a variety of journals. *Experimental Agriculture* (Oxford University Press, U.K.); *Tropical Agriculture* (Trinidad); *International Agriculture* (Agraria Press, U.K.).; *Journal of Soil and Water Conservation* (Ankeny, Iowa, U.S.A); *Indian Journal of Soil and Water Conservation* (Dehra Dun); *Soil and Water* (National Water and Soil Conservation Authority, Wellington, New Zealand) all contain useful reading for the watershed manager.

Forestry journals and range management journals occasionally contain tropical watershed papers, but when studies of watershed effects of afforestation or of grazing management are published they are usually in the hydrology journals. The journal *Agroforestry* was recently begun by Elsevier, Amsterdam. There are several journals of ecology, but these are specialist literature more concerned with habitat description than with watershed management. The research literature is abstracted and computerized, but few papers on watershed policy and development experience are included.

Newsletters

Newsletters are valuable as sources of "current awareness" information. The World Association of Soil and Water Conservation issues a highly relevant newsletter. The association is sponsored by the Soil Conservation Society of America (Ankeny, Iowa, U.S.A). The International Irrigation Information Center (see above) publishes a newsletter, *Irrinews,* that contains useful book reviews and current news items. The International Council for Research on Agroforestry (PO Box 30677, Nairobi, Kenya) issues a newsletter that has much direct relevance to watershed management. The International Rice Research Institute's newsletter (PO Box 933, Manila, Philippines) is useful for watersheds in which rice is a major crop. For drier climates, watershed managers will also find items of interest in the newsletter on research progress at the International Crops Research Institute for the Semi-Arid Tropics (PO Patancheru, A.P. 502324, India). The Australian Centre for International Agricultural Research has begun a newsletter that is concerned with Southeast Asia and the Pacific (GPO Box 1571, Canberra 2601, Australia). The East-West Center (Honolulu, Hawaii, U.S.A) produces a lively bulletin for the humid tropics, the *Asia Pacific Forest Watershed Newsletter.*

Bibliographies

Recent bibliographies are:

Blackie, J.R.; Ford, E.D.; Horne, J.E.M.; Kinsman, D.J.J.; Last, F.T.; and Moorhouse, P. 1980. *Environmental effects of deforestation.* Occasional Publication no. 10. Ambleside, U.K: Freshwater Biological Association

Cruz, R.V.; Birones, L.; and Hufschmidt, M.M. 1984. *Water resources management in Asia: A selective bibliography with introductory essays.* Honolulu: East-West Center.

Lundgren, B., and Samuelson, A. M. 1975. *Land use in Kenya and Tanzania: A bibliography.* Stockholm: Royal College of Forestry.

Recommended reading

Brown, L.R., and Wolf, E.C. 1984. *Soil erosion: Quiet crisis in the world economy.* Worldwatch Paper 60. Washington D.C.: Worldwatch Institute.

FAO. 1976. *Hydrological techniques for upstream conservation.* FAO Conservation Guide no. 2. Rome.

FAO. 1977. *Mountain forest roads and harvesting.* Forestry Papers no. 14. Rome.
FAO. 1977. *Conservation in arid and semi-arid zones.* FAO Conservation Guide no. 3. Rome.
FAO. 1977. *Guidelines for watershed management.* FAO Conservation Guide no. 1. Rome.
German Agency for Technical Cooperation. 1975. *Manual on reforestation and erosion control for the Philippines.* Eschborn, W. Germany.
Hamilton, L.S. (ed.). 1983. *Forest and watershed development and conservation in Asia and the Pacific.* Boulder, Colo.: Westview Press.
Hillel, Daniel. 1982. *Negev: Land, water and life in a desert environment.* New York: Praeger.
Hudson, N.W. 1975. *Field engineering for agricultural development.* Oxford: Clarendon Press.
Hudson, N.W. 1981. *Soil conservation.* London: Batsford Academic.
Jackson, I.J. 1977. *Climate, water and agriculture in the tropics.* London: Longman.
Man and the Biosphere Programme. 1975. *The Sahel: Ecological approaches to land use.* MAB Technical Notes. Paris: UNESCO Press.
National Academy of Sciences. 1974. *More water for arid lands.* Washington D.C.: National Academy Press.
National Academy of Sciences. 1980. *Firewood crops.* Washington D.C.: National Academy Press.
National Academy of Sciences. 1984. *Agroforestry in the West African Sahel.* Washington D.C.: National Academy Press.
Office of Technology Assessment. 1984. *Technologies to sustain tropical forest resources.* Washington D.C.: Congress of the U.S.
Penman, H.L. 1963. *Vegetation and hydrology.* Technical Communication no. 53. Farnam Royal, U.K.: Commonwealth Agricultural Bureaux.
Pereira, H.C. 1973. *Land use and water resources.* Cambridge: Cambridge University Press.
Sanchez, P.A. 1976. *Properties and management of soils in the tropics.* New York: Wiley.
Schultz, F.E., ed. 1980. *Land reclamation and water management.* Wageningen, Netherlands: International Institute for Land Reclamation and Improvement.
Singh, G.; Venkataramanan, C.; and Sastry, G. 1981. *Manual of soil and water conservation practices in India.* ICAR Bulletin no. T./13D.10. Dehra Dun: Indian Council of Agricultural Research.
UNESCO/UNDP/FAO. 1979. *Tropical grazing land ecosystems.* Paris: UNESCO, Natural Resources Research Dept.

U.S. Department of Agriculture. 1979. *Field manual for research in agricultural hydrology.* Agricultural Handbook no. 224. Washington D.C.

Watson, G.A. 1980. *A study of the tree-crop farming systems of the humid tropics.* Agriculture Technical Note no. 2. Washington, D.C.: World Bank

World Resources Institute. 1985. *Tropical forests: A call for action.* Washington. D.C.

Wortman, S., and Cummings, R.W., Jr. 1978. *To feed this world.* Baltimore, Md.: John Hopkins University Press.

Index